3rd International Meeting on Silicene 2013

(IMS-3)

Journal of Physics: Conference Series Volume 491

Istres, France
10-14 June 2013

Editors:

Abdelkader Kara
Jean Louis Lemaire
Hanna Enriquez
Hamid Oughaddou

ISBN: 978-1-63266-716-8
ISSN: 1742-6588

Printed from e-media with permission by:

Curran Associates, Inc.
57 Morehouse Lane
Red Hook, NY 12571

Some format issues inherent in the e-media version may also appear in this print version.

Printed by Curran Associates, Inc. (2014)

For permission requests, please contact the Institute of Physics
at the address below.

Institute of Physics
Dirac House, Temple Back
Bristol BS1 6BE UK

Phone: 44 1 17 929 7481
Fax: 44 1 17 920 0979

techtracking@iop.org

Additional copies of this publication are available from:

Curran Associates, Inc.
57 Morehouse Lane
Red Hook, NY 12571 USA
Phone: 845-758-0400
Fax: 845-758-2634
Email: curran@proceedings.com
Web: www.proceedings.com

TABLE OF CONTENTS

Journal of Physics: Conference Series **491** (2014) 011001 doi:10.1088/1742-6596/491/1/011001

3rd International Meeting on Silicene (IMS -3)

June 12th 2013, Istres, France

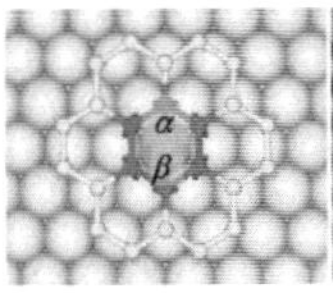

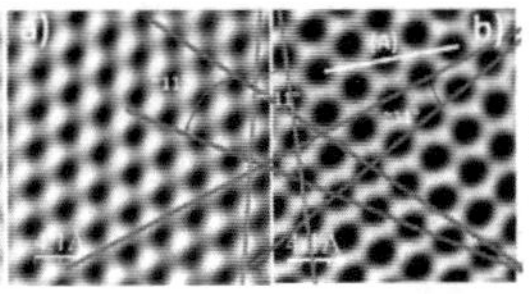

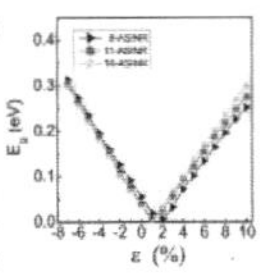

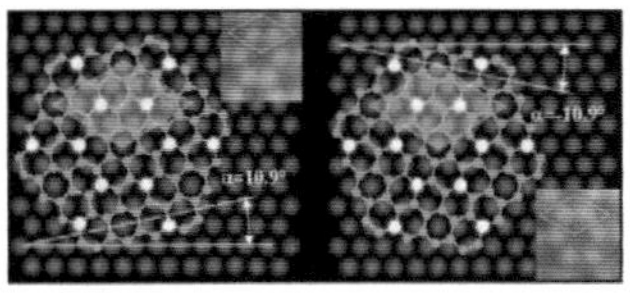

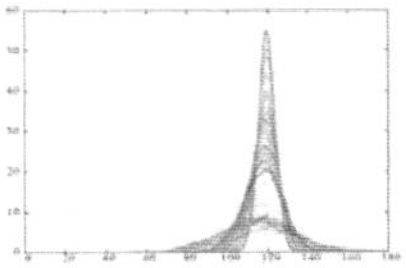

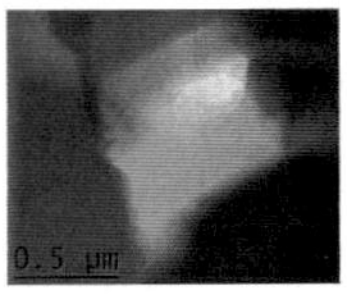

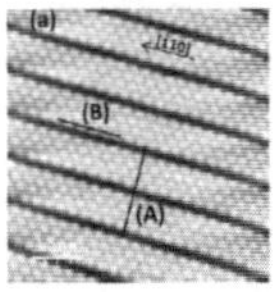

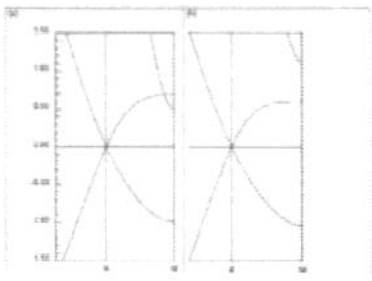

Editorial committee

Professor Abdelkader Kara

Professor Jean Louis Lemaire

Dr Hanna Enriquez

Professor Hamid Oughaddou

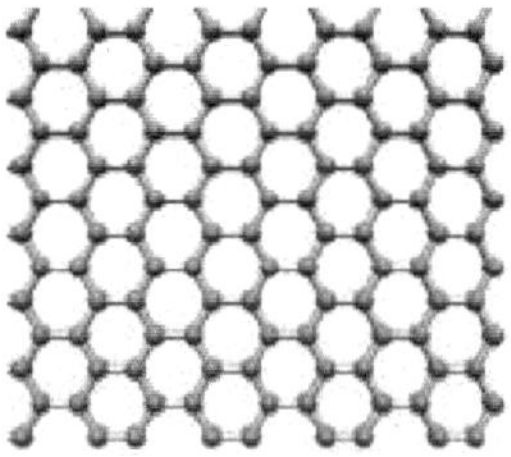

Preface

Silicene, the new supernova material

Silicon is formed in a large amount during supernova nucleo-synthesis and it is the 8th most common chemical element in the universe.

Today modern electronics make large use of bulk silicon, which has consequently an extremely large impact on the world industry and economy. The need for more powerful, faster and less energy-consuming integrated circuits requires in the future the use of nanotechnologies. The ultimate step concerning silicon is silicene (the 2D silicon-based analogue of graphene). This material is of paramount importance as it requires the use of the same technologies and production lines as silicon.

Even if theoreticians had predicted its possible existence, it is only in 2010 that a team of pioneers from CINAM-France, ISMO-France and UCF-USA has presented for the first time the experimental evidence of the formation of silicene. Since then research is exploding (like a supernova!) both on the experimental and theoretical sides, with the main aim of replacing bulk silicon with this potentially revolutionary material. However, before any possible industrial use, it has first to be prepared or synthesized in various stripe and sheet shapes on insulating surfaces on which its physical and chemical properties have to be analyzed in detail. A second step is its chemical functionalization through various dopants to achieve different tasks, expected or not yet imagined.

Chemists and physicists, experimentalists and theoreticians are involved in this thrilling work. A wide array of techniques, from the subtle chemistry reaction networks, to all those of experimental surface science (from synchrotron radiation to scanning tunneling microscopy) as well as those of theoretical chemistry (from *ab initio* to density functional theory calculations) are involved.

Big progress has been made since 2010 and the success of this third International Meeting on Silicene (IMS-3) is demonstrated by the present proceedings.

Historical summary

Every two years, the STARM (science, technologie avancée et recherche pour la Mediterranée, http://www.starm.emcmre.org/) society is organizing an international conference entitled Euro-Mediterranean Conference on Materials and Renewable Energies (EMCMRE, http://www.emcmre.org/) in countries across the Mediterranean Sea.

It is in this framework that an international meeting dedicated to silicene is organized simultaneously since 2010:

1st International Meeting of Silicene (IMS-1), Safi, Morocco, 2010
2nd International Meeting of Silicene (IMS-2), Marrakech, Morocco, 2011
3rd International Meeting of Silicene (IMS-3), Istres-Marseille, France, 2013

Peer review statement

All papers published in this volume of *Journal of Physics: Conference Series* have been peer reviewed through processes administered by the proceedings Editors. Reviews were conducted by expert referees to the professional and scientific standards expected of a proceedings journal published by IOP Publishing.

Silicene on Ag(111) : domains and local defects of the observed superstructures

Haik Jamgotchian[1], Yann Colignon[2], Bénédicte Ealet [1], Bence Parditka[2,3], Jean-Yves. Hoarau[1], Christophe Girardeaux[2], Bernard Aufray[1,a] and Jean-Paul. Bibérian[1]

[1] Aix-Marseille Université, CNRS, CINaM, UMR 7325, 13288 Marseille, France
[2] Aix-Marseille Université, CNRS, IM2NP, UMR 7334, 13397 Marseille, France
[3] Department of Solid State Physics, University of Debrecen, H-4026 Debrecen, Bem ter 18/b., Hungary

E-mail: [a]aufray@cinam.univ-mrs.fr

Abstract. On Ag(111) substrate, the growth of a silicene layer (the equivalent of graphene for silicon) gives rise to four main superstructures due to the epitaxy between the silicene layer (honeycomb structure) and the Ag(111) substrate. Depending on both the substrate temperature and the deposition rate, the following superstructures were observed: (4×4), $(\sqrt{13} \times \sqrt{13})R13.9^\circ$, $(2\sqrt{3} \times 2\sqrt{3})R30^\circ$ and $(\sqrt{7} \times \sqrt{7})R19.1^\circ$. Each one corresponds to a specific rotation of the silicene layer with respect to the silver substrate. In this paper we show and discuss for each superstructure all the expected equivalent domains due to the symmetry properties of both the silicene honeycomb structure and the trigonal silver surface. Finally, we show, from STM images of the $(2\sqrt{3} \times 2\sqrt{3})R30^\circ$ superstructure, the role played by periodic local defects on the average direction of the superstructures.

1. Introduction

After the publication of three important papers showing a possible synthesis of silicene (graphene-like silicon monolayers) on the silver substrates (Ag(100)[1] Ag(110)[2] and Ag(111)[3] faces), many groups all around the world started to work on this system [4]. Recently and especially on the Ag(111) face, the growth of silicene layers has been confirmed even if, different atomic models were proposed [5, 6, 7, 8, 9, 10, 11, 12, 13, 14]. At least four superstructures have been observed and/or predicted due to the epitaxy between the silicene layer (honeycomb structure) and the Ag(111) substrate (trigonal structure). Depending on the substrate temperature and the deposition rate, different superstructures are observed. A (4x4) superstructure is obtained if the substrate temperature is maintained between 150 and $180°C$ and with a relatively slow deposition rate (0.5 ML / h). A $(2\sqrt{3} \times 2\sqrt{3})R30^\circ$ is obtained with a substrate temperature above $280°C$ and with a higher deposition rate (5 ML/h). In between these two experimental growth conditions, a mixture of these two superstructures plus a $(\sqrt{13} \times \sqrt{13})R13.9^\circ$ superstructure is obtained. As expected, the (4×4) superstructure is due to the perfect match between four silver inter-atomic distances (1,156 nm), and three silicon hexagons (1,152 nm), both structures oriented with the same [110] directions. We have previously shown that starting from the (4×4) superstructure, a rotation α of the silicene layer relatively to the silver substrate gives rise successively to a first $(\sqrt{13} \times \sqrt{13})R13.9^\circ$

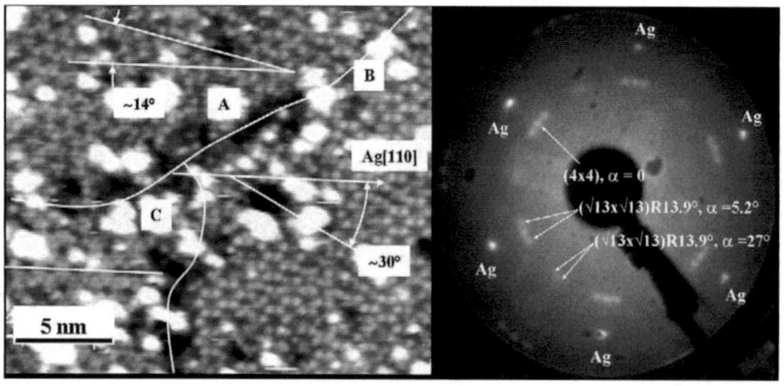

Figure 1. a) STM image of a mixture of superstructures. ($V = -1.1V$, $I = 1.2nA$) b) Corresponding LEED pattern showing: $(\sqrt{13} \times \sqrt{13})R13.9°$, domain A, $(2\sqrt{3} \times 2\sqrt{3})R30°$, domain B and (4×4) domain C .

($\alpha = 5.2°$), a $(2\sqrt{3} \times 2\sqrt{3})R30°$ ($\alpha = 10.9°$), a $(\sqrt{7} \times \sqrt{7})R19.1°$ ($\alpha = 19.1°$), and finally a second $(\sqrt{13} \times \sqrt{13})R13.9°$ ($\alpha = 27°$) structure [6]. In this paper we show and discuss for each superstructure all the expected equivalent domains due to the symmetry properties of both the silicene honeycomb structure and the trigonal silver surface. Finally, we show from STM images of the $(2\sqrt{3} \times 2\sqrt{3})R30°$ superstructure, the role played by periodic local defects on the average direction of the superstructures.

2. The five main superstructures and their associated domains.

From a general point of view, the LEED pattern of a given superstructure, with respect to the substrate, reveals two important information: the first one is the over-periodicity (i.e. a length) and the second one is the orientation of the superstructure (i.e. a direction). As a consequence, when different superstructures are present simultaneously on the same surface, both information (periodicity and direction) are used in order to identify the different superstructures observed by Scanning Tunneling Microscopy (STM) [15]. When the silicene were grown on Ag(111) then we generally observed by STM and LEED the coexistence of different superstructures which are difficult to identify (see Figure 1 (a)). Since the silver sample is always in the same orientation before and after the silicene growth, we used the information extracted from LEED (direction and periodicity) to identify each superstructure. This is how we had shown [6], and later confirmed [11] that morphologies of the silicene layer reflects the position of each silicon atom with respect to the silver ones: in other words when a silicon atom is situated on top of a silver atom, it appears always higher than those located in threefold sites or in bridge positions, as shown schematically in figure 2. All the STM images, which are presented in this paper, have been obtained after the same experimental protocol explained in detail in [6].

3. The single domain (4×4) superstructure.

This superstructure is obtained thanks to the perfect match amongst four nearest neighbour silver distances (1.156nm) and three hexagonal cells of the (111) surface of silicon (1.152nm). Due to the superposition of both axes of symmetry between the silver and the silicene planes ($\alpha = 0°$), there is only one domain. As shown in figure 3, the ball model starting with hexagons of silicon atoms around a silver atom explains the STM image: the silicon atoms located close to the top position of silver atoms correspond to the intense spots of the STM image. DFT calculations have confirmed the first model, showing that the density of states is located around the silicon atoms [8].

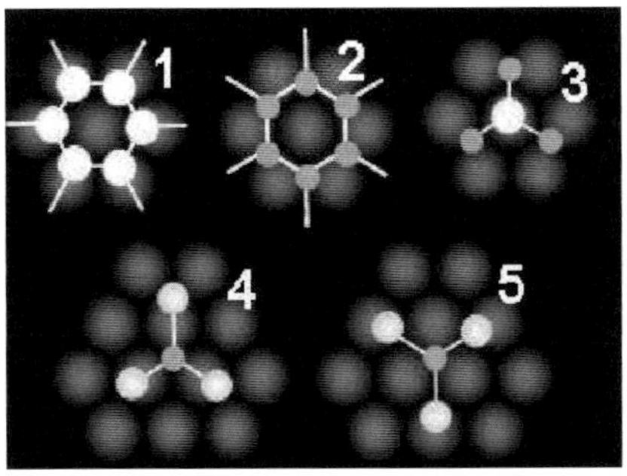

Figure 2. Ball models of adsorption sites of silicon on silver used to interpret all the STM images. 1) six silicon atoms close to top positions around a silver atom. 2) six silicon atoms close to three fold positions around a silver atom. 3) One silicon atom on top position and three silicon atoms in bridge positions. 4) One silicon atom in a three-fold site and three silicon atoms close to top positions. 5) Mirror image of 4)

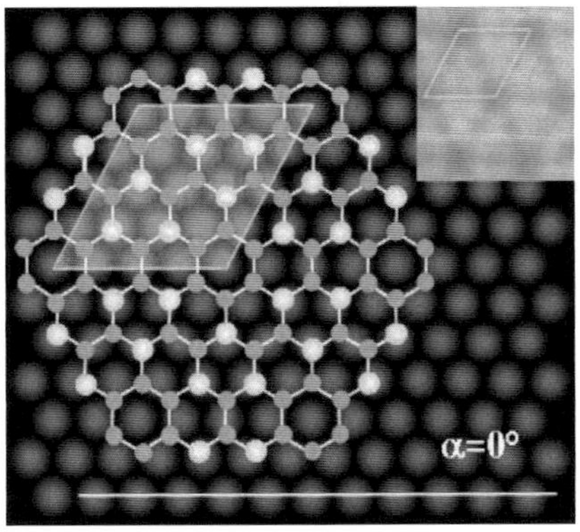

Figure 3. Ball models of the (4×4) superstructure and the corresponding STM image.

4. The two $(\sqrt{13} \times \sqrt{13})R13.9°$ superstructures and four domains

The LEED pattern of the $(\sqrt{13} \times \sqrt{13})R13.9°$ shows two directions for the superstructure corresponding to two angles: $\alpha = 5.2°$, and $\alpha = 27°$ as shown in figure 1 b. The corresponding STM image on figure 4 shows the two different $(\sqrt{13} \times \sqrt{13})R13.9°$ structures .

For: $\alpha = 5.2°$, the superstructure not being on a silver axis of symmetry, there are two mirror domains rotated: $+13.9°$ and $-13.9°$ with respect to the Ag[110] direction. For the sake of simplicity, only one orientation: $+13.9°$ is discussed. The ball model and the corresponding STM image are shown in figure 5 a. Here again, the intense spots of the STM image correspond to silicon atoms located on top of silver atoms.

For: $\alpha = 27°$, the superstructure corresponds to a mirror structure with respect to the Si[110] direction. For this structure there are also two domains rotated: $+13.9°$ and $-13.9°$ with respect to the Ag[110] direction. The ball model for $+13.9°$ and the corresponding STM image are shown in figure 5 b. Here there is no silicon atom on top of silver atoms. Instead, we observe intense areas corresponding to groups of silicon atoms.

Finally, there is a total of four domains for this superstructure. This is a unique case where the same superstructure $(\sqrt{13} \times \sqrt{13})R13.9°$ has two different atomic compositions in the unit cell. The LEED pattern corresponding to each one of these structures shows the same position of the diffraction spots, but with very different intensities (Figure 1 b). The intense spots of each of these superstructures correspond to the silicene unit cell: distance and angle between two silicon hexagons [15].

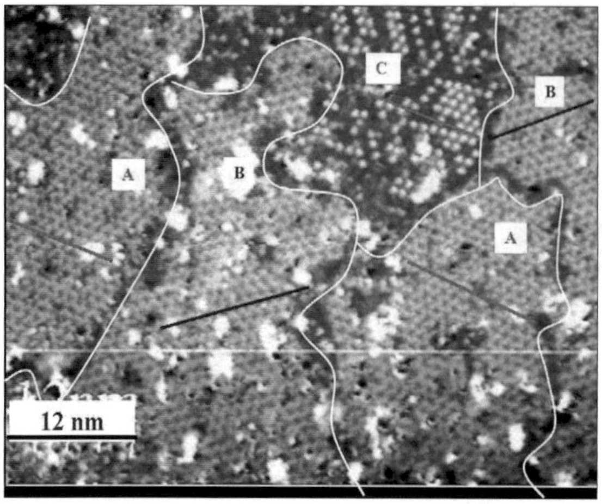

Figure 4. STM image showing a mixture of the two different $(\sqrt{13}\times\sqrt{13})R13.9°$ superstructures ($V = -1.4V$, $I = 800pA$). Domains A and B correspond to the same superstructures ($\alpha = 5.2°$ and $\alpha = -5.2°$respectively). Domain C is the other type of superstructure ($\alpha = 27°$). All domains are separated by grain boundaries.

Figure 5. Ball model of the two different $(\sqrt{13}\times\sqrt{13})R13.9°$ superstructures with (a) $\alpha = 5.2°$ and (b) $\alpha = 27°$

5. The two domains $(2\sqrt{3}\times 2\sqrt{3})R30°$ superstructure

This superstructure is obtained when the silicene layer is rotated by an angle, $\alpha = 10.9°$, with respect of the silver substrate (Figure 6). The direction of the $(2\sqrt{3}\times 2\sqrt{3})R30°$superstructure is on a silver symmetry axis, so we expect only two domains due to the mirror structure with respect to the Si[110] direction as it is shown on the figure 6. The two domains exist on the Ag surface but cannot be separated by STM since they form the same protrusions structure as it is shown on the ball model of figure 6.

6. The two domains of the $(\sqrt{7}\times\sqrt{7})R19.1°$ superstructure.

This superstructure was predicted [6] based on the good match between the silicene layer and the silver lattice. This superstructure has been later observed by Chiappe et al. [11]. The STM morphology shows big protrusions forming a trigonal network with the periodicity of 0.76 nm rotated 19.1° with respect to the Ag[110] direction. This superstructure not being on a silver axis of symmetry, two domains rotated 19.1° and −19.1° are expected. As shown on figure 7,

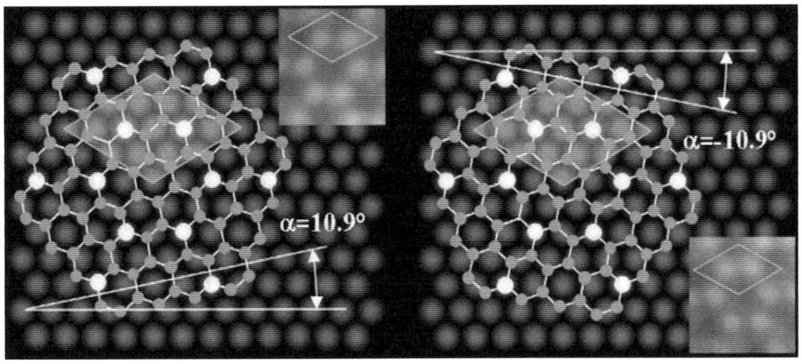

Figure 6. The two models of the $(2\sqrt{3}\times 2\sqrt{3})R30°$ superstructure: a) $\alpha = +10.9°$ b) $\alpha = -10.9°$

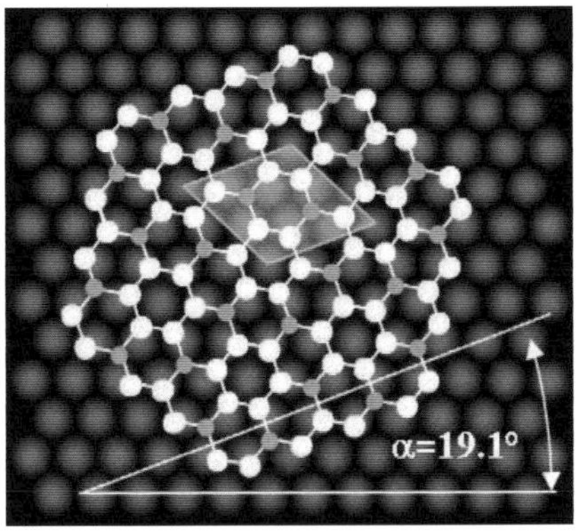

Figure 7. Ball model of the $(\sqrt{7}\times\sqrt{7})R19.1°$ superstructure

a model similar to the $(\sqrt{13}\times\sqrt{13})R13.9°$ superstructure with 6 silicon atoms around a silver atom is proposed. For this model also, three first silicon atoms are also situated close to the top of silver atoms giving rise to the big protrusion observed by STM [11].

7. Effects of local defects

As shown on the filled states STM images on figure 8, the direction of the $(2\sqrt{3}\times 2\sqrt{3})R30°$ superstructure is not exactly at an angle of 30°. This observation is explained by regular periodic defects formed around a perfect domain as it is highlighted on the figure 8, giving rise to the observed average slight rotation shift, in this case about 4°. The value of the rotation shift can change depending on the size of the $(2\sqrt{3}\times 2\sqrt{3})R30°$ domains: the larger the domains, the smaller the angle shifts. In addition, this analysis could also explain the Moiré pattern observed by Feng et al. [5].

Interestingly figure 9 shows two domains of $(2\sqrt{3}\times 2\sqrt{3})R30°$ with defects both close to 30° one rotationally shifted positively, and the other negatively, separated by a grain boundary. Each domain corresponds to a rotation of the silicene layer, $\alpha = +10.9°$ and $\alpha = -10.9°$. The approximate 7° disorientation highlighted in figure 9 corresponds to approximately two times 3.5° : i.e. 3.5° for $\alpha = +10.9°$ and $-3.5°$ for $\alpha = -10.9°$. When alpha is +10.9° the local defects deform the structure towards the $(\sqrt{13}\times\sqrt{13})R13.9°$structure with the silicene layer rotated by +5.2°. Similarly when alpha is −10.9° the local defects deform the structure towards the $(\sqrt{13}\times\sqrt{13})R13.9°$ structure with the silicene layer rotated by −5.2°.

In this case local defects permits to distinguish the two different $(2\sqrt{3}\times 2\sqrt{3})R30°$ structures that are undistinguishable by STM as shown on the ball models of figure 6.

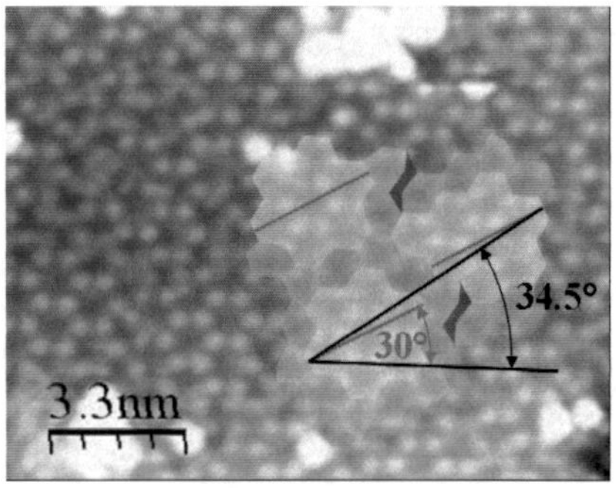

Figure 8. STM image of the $(2\sqrt{3} \times 2\sqrt{3})R30°$ superstructure showing the rotation shift of the average orientation resulting from local defects between perfect domains. Highlighted the domains and the defects. $(V = -1.7V, I = 1.1nA)$

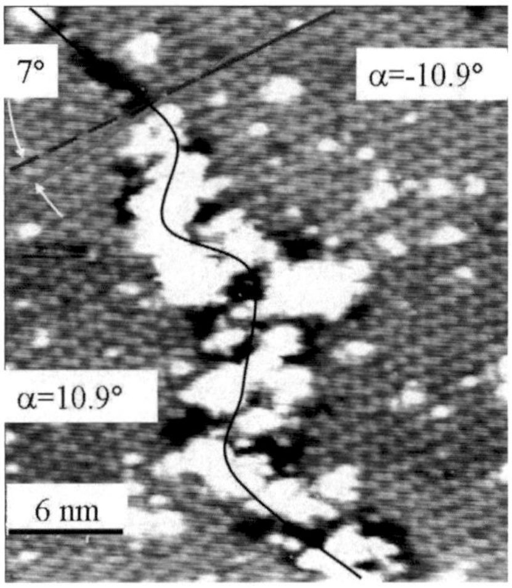

Figure 9. STM image showing two $(2\sqrt{3} \times 2\sqrt{3})R30°$ domains with local defects giving rise to a grain boundary $(V = -2.0V,\ I = 900pA)$

8. Conclusion

In this paper we have shown each one of the five superstructures of silicene on silver (111): the (4x4), the two different $(\sqrt{13} \times \sqrt{13})R13.9°$, the $(2\sqrt{3} \times 2\sqrt{3})R30°$ and the $(\sqrt{7} \times \sqrt{7})R19.1°$, We explained the LEED and STM observations by ball models. In the case of the $(2\sqrt{3} \times 2\sqrt{3})R30°$ superstructure, we show that local defects located around perfect $(2\sqrt{3} \times 2\sqrt{3})R30°$ areas justify the average orientation shift observed in the $(2\sqrt{3} \times 2\sqrt{3})R30°$.

References

[1] Léandri C, Oughaddou H, Aufray B, Gay J.M., Le Lay G, Ranguis A et al 2007 *Surface Science.* **601** 262
[2] Aufray B, Kara A, Vizzini S, Oughaddou H, Léandri C, Ealet B and Le Lay G 2010 *App. Phys. Lett.* **96** 183102
[3] Lalmi B, Oughaddou H, Enriquez H, Kara A, Vizzini S, Ealet B and Aufray B 2010 *Appl. Phys. Lett.* **97** 223109
[4] Kara A, Enriquez H, Seitsonen A.P, Lew Yan Voon,L.C, Vizzini S, Aufray B et al 2012 *Surface Science Reports* **67** 1
[5] Feng B, Ding Z, Meng S, Yao Y, He X, Cheng P, Chen L and Wu K 2012 *Nano Lett.* **12** 3507
[6] Jamgotchian H, Colignon Y, Hamzaoui N, Ealet B, Hoarau J Y, Aufray B and Bibérian J P 2012 *J. Phys. Cond. Mat.* **24** 172001

[7] Lin C, Arafune R, Kawahara K, Tsukahara N, Minamitani E, Kim Y, Takagi N and Kawai M 2012 *App. Phys. Exp.* **5** 045802
[8] Vogt P, De Padova P, Quaresima C, Avila J, Frantzeskakis E, Asensio M.C et al 2012 *Phys. Rev. Lett.* **108** 155501
[9] Arafune R, Lin C-L, Kawahara K, Tsukahara N, Minamitani E, Kim Y et al 2013 *Surf. Sci.* **608** 297
[10] Chen L, Liu C-C, Feng B, He X, Cheng P, Ding Z et al 2012 *Phys. Rev. Lett.* **109** 056804
[11] Chiappe D, Grazianetti C, Tallarida G, Fanciulli M and Molle A 2012 *Advanced Materials* **24** 5088
[12] Enriquez H, Vizzini S, Kara A, Lalmi B and Oughaddou H 2012 *J. Phys.Condens. Matter* **24** 314211
[13] Gao J and Zhao J 2012 *Sci. Rep.* **2** 861
[14] Kaltsas D, Tsetseris L and Dimoulas A 2012 *J. Phys.Condens. Matter* **24** 442001
[15] Perdereau J and Rhead G.E 1971 *Surf. Sci.* **24** 555

Atomic structure of silicene nanoribbons on Ag(110)

Mohammed Rachid Tchalala[1,3], Hanna Enriquez[1], Andrew J. Mayne[1] Abdelkader Kara[2], Gérald Dujardin[1], Mustapha Ait Ali[3] and Hamid Oughaddou[1,4]

[1] Institut des Sciences Moléculaires d'Orsay, ISMO-CNRS, Bâtiment 210, Université Paris-Sud, 91405 Orsay, France

[2] Department of Physics, University of Central Florida, Orlando, Florida, 32816 USA

[3] Laboratoire de Chimie de Coordination et Catalyse, Département de Chimie, Faculté des Sciences-Semllia, Université Cadi Ayyad, Marrakech, 40001, Morocco

[4] Département de Physique, Bâtiment Neuville II, 5 Mail Gay-Lussac, 95031 Neuville sur Oise, Cergy-Pontoise Cedex, France

Abstract. The growth of silicene nano-ribbons (NRs) on Ag(110) substrate is re-investigated using scanning tunneling microscopy (STM) and low energy electron diffraction (LEED). Deposition of one silicon monolayer at 230°C induces the formation of one-dimensional 1.6 nm wide silicene nanoribbons into a well-ordered compact array with a nanometer-scale pitch of just 2 nm. Based on the STM analysis we derived an atomic model of the silicene nanoribbons (NRs) where they are substantially buckled, and quantum confinement of the electrons in the NRs contribute to electronic density of states.

1. Introduction

Substantial efforts have been focused recently on the electronic properties of two-dimensional (2D) materials in the perspective of their integration new devices [1]. Given the impact of graphene over the last few years [2], the isomorphic silicene [3] and germanene [4] are now considered to be promising novel materials with the advantage of being compatible with existing semiconductor technology [5]. Theoretical investigations of free-standing silicene show that these two novel materials have electronic structures with band structures presenting a linear crossing at the K and K' points of the surface Brillouin zone indicative of semi-metallic character, and massless Dirac fermion character. These result in similar properties such as a large charge carrier mobility and Hall effect [4,6-8]. For the moment, the primary method for growing silicene and germanene is on solid substrates, in particular silver: Ge/Ag [9-12] and Si/Ag [3, 13-15] and recently on gold [16] and iridium [17]. On Ag(111), silicene forms a continuous sheet [15], with at least three distinct ordered phases [18-21], depending on the deposition conditions. Free-standing silicon sheets only exist as silicon nanotubes [22] or multilayer silicene from chemical methods [23]. Now, nanoribbons are very interesting because of their potential for 1D dispersion effects with directional conductivity. A perfectly well defined edge structure is required and the coherent electron confinement will lead to observable effects such as quantum interference channeling at their edges [24]. To achieve this, graphene nanoribbons have to be synthesized chemically [25, 26], and Bi nanolines on Si(001) require a high formation temperature (600°) [27], while Si nanowires on SiC did not show any 1D conductivity [28]. On the Ag(110) surface, one-dimensional silicene nanoribbons (NRs) [13] can be grown. The isolated NRs show a low reactivity to molecular oxygen; oxidation occurs only at the Si NRs terminations and develops along the [-110] direction indicating that their edges are not reactive to molecular oxygen [29]. In response to this remarkable observation, we present Scanning Tunneling Microscopy images of a compact parallel assembly of silicene nanoribbons on Ag(110). The NRs are 1.6 nm wide with a well-defined edge structure and an electronic contribution is visible in the STM images taken at different voltages.

2. Experiments

The standard tools for surface preparation and characterization were used: an ion gun for surface cleaning, a low energy electron diffraction (LEED) instrument, and a scanning tunneling microscope (STM) for surface characterization at the atomic scale. The STM experiments were performed at room temperature (RT). The Ag(110) sample was cleaned by several cycles of sputtering ($600eV Ar^+$ ions, $P = 5 \cdot 10^{-5} mbar$) followed by annealing at 400°C until a sharp p(1x1) pattern was obtained. Silicon was evaporated by a direct current heating of a piece of Si wafer, and deposited onto the Ag(110) surface held at 230°C.

3. Results and discussion

Figure 1a shows the bright sharp (1x1) LEED pattern of the clean bare Ag(110) surface. The deposition of silicon on the silver substrate is done at a constant flux and the LEED patterns are regularly monitored. We found that during silicon deposition on the Ag(110) substrate held at 230°C , a (2x5) LEED pattern appears and becomes brighter as we approach a Si coverage of 1 monolayer (ML), as clearly seen in Figure 1b. The (x2) periodicity is along the $[\bar{1}10]^*$ while the spots along the $[001]^*$ direction show a (x5) periodicity relative to the bare substrate. To guide the eye, the (x5) and the (x2) spots are highlighted by white circles. This clear LEED pattern is a signature that 1ML of silicon forms an extended periodic superstructure on Ag(110).

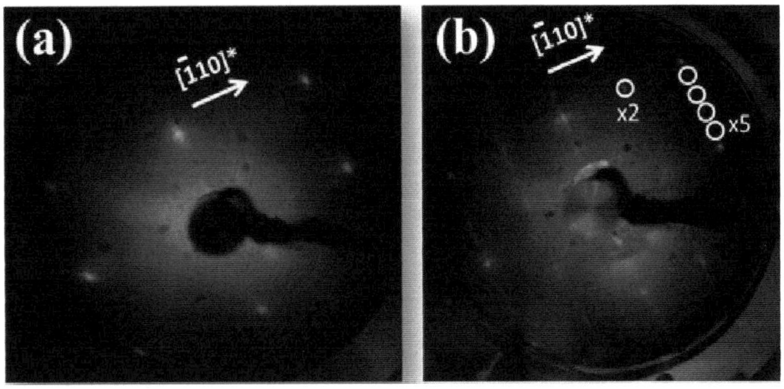

Figure 1. LEED patterns recorded at $E_p = 50eV$ corresponding to: a) the (1x1) reconstruction of the bare Ag(110) surface, b) the (2x5) superstructure obtained after the deposition of 1 Si ML on Ag(110) held at 230° C. The (x2) and (x5) spots are highlighted by white circles.

We performed STM measurements at several imaging conditions to probe atomic structure of this (2x5) silicon superstructure in more detail. The (2x5) superstructure STM image (Figure 2a), obtained at a sample voltage V = -80 mV and a constant current of 2.2 nA, shows a well-ordered and compact array of self-assembled silicene nanoribbons (NRs). All these silicene NRs are aligned along the [-110] direction of the bare Ag(110). By scanning large areas of the surface, we noted that their length is limited only by the size of the terraces. In order to quantitatively determine the structural characteristics of the silicon NRs, several line scans were analyzed and shown in Figures 2b and 2c. The line scans reveal NRs with a width of 1.6 nm, corresponding to $4a_{Ag}$ where a_{Ag}=0.409 nm, and they are evenly spaced with a separation of 2.0 nm corresponding to a (x5) periodicity along the Ag [100] direction (Figure 2b). In the STM image, the NRs are composed of an ordered array of protrusions (bright spots) with a spacing along the Ag [-110] direction corresponding to a (x2) periodicity (Figure 2c). The distance between two neighbor protrusions along the Ag $[\bar{1}10]$ direction is equal to 0.58 nm corresponding to $2a_{[\bar{1}10]}$, in good agreement with the (x2) periodicity shown in the LEED pattern along the Ag [-110] direction.

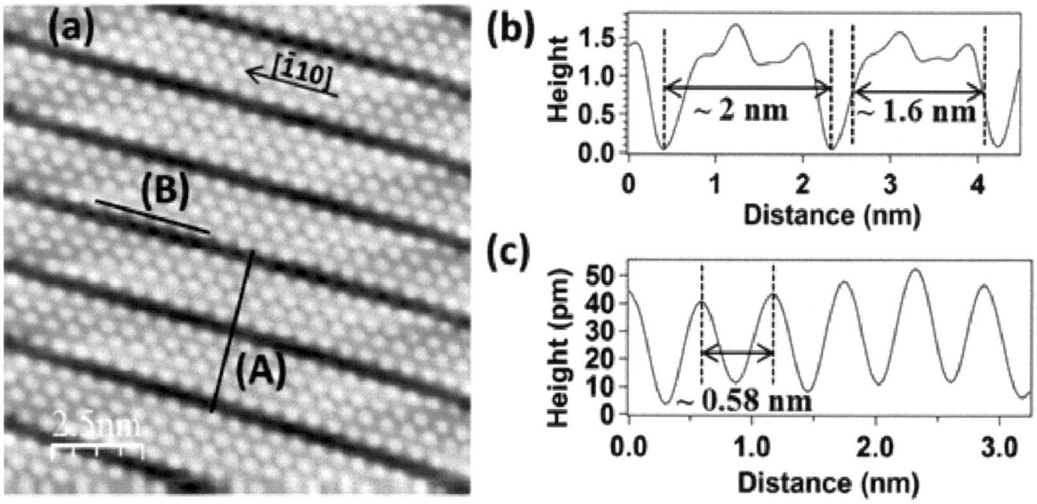

Figure 2. a) STM image of silicene NRs grown at 230° C on a Ag(110) surface (12.5x12.5 nm^2, V = -80 mV, I = 2.2 nA), b) line scan A across the NR showing the (x5) periodicity, c) line scan B along the NR showing the (x2) periodicity.

In Figure 2, the lateral size of the protrusions within the NRs is too large to correspond to single silicon atoms. So the silicon NRs were imaged under different conditions from those of Figure 2. At a surface voltage of -0.2V and a constant current of 1.9 nA, we observed a honeycomb arrangement (highlighted in red in Figure 3a) indicative of a quasi-atomic resolution. The pattern is clearly quite different from that in Figure 2. An analysis of the line scans shows that the distance between the nearest-neighbor protrusions forming this honeycomb structure (red hexagons) is equal to 0.335 nm. This distance is much larger than the expected nearest-neighbor Si-Si distance of 0.235 nm, indicating that the protrusions observed in the STM image do not correspond to neighboring Si atoms. Based on these experimental observations, we derived a structural model, which is displayed in Figure 3b. In this model we assign the protrusions observed in STM to the two second-neighbor silicon atoms of an armchair-terminated silicene nanoribbon. Now, the 0.58 nm wide hexagons are commensurate with the (x2) periodicity along the Ag direction, giving a measured lateral distance of 0.335 nm between the two second nearest silicon atoms. This results in an in-plane lateral separation between nearest-neighbor Si atoms of 0.193 nm. There are two explanations why not all the silicon atoms in the NRs are visible in the STM images. The first is that the silicon atoms in the nanoribbon are substantially buckled, as recent studies on silicene have reported [21, 29] with a buckling between 0.04 and 0.12 nm. The reported Si-Si bond length in silicene sheets varies between 0.20 and 0.25 nm, which implies that the projected in-plane Si-Si distances are close to the 0.193 nm value measured in this study and previously [15, 29].

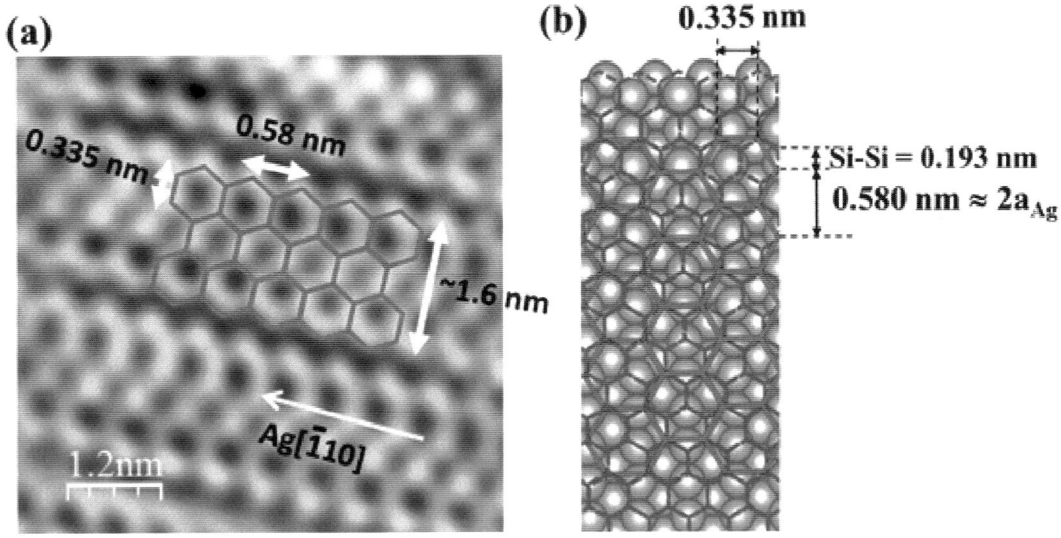

Figure 3. a) Atomically resolved filled-states STM image of silicene NRs grown at 230° C on a Ag(110) surface (6x6 nm^2, V = -0.2 V, I = 1.9 nA). b) Proposed model where the grey and yellow atoms correspond to the first and the second silver layers, respectively. The blue honeycomb structure corresponds to the internal structure of silicene NRs while the red one corresponds to the structure observed in the STM image.

The second explanation is an electronic contribution to the periodic structure in the STM images, which is controlled by the interaction between the atoms of the silicene and the rows of underlying Ag atoms. The graphene and silicene honeycomb structure has a Bravais lattice containing two atoms which leads to an inequivalence between the sub-lattices "a" and "b". Inspection of Figure 3a immediately shows that the red hexagons occupy 2/3 of the atoms of one lattice and the central "chains" are aligned between the Ag atom rows. Recently, theoretical calculations indicate that electron transport occurs coherently via atoms of the same sub-lattice [31]. Also, the 1D nanoribbon confines the 2D electron gas giving rise to electron scattering at armchair edges with the formation of quantum interferences localized on the bonds [25]. These two related effects indicate that the electron density is channeled along the nanoribbons.

4. Conclusion

In conclusion, we have presented a detailed experimental study of the epitaxial growth of silicon on Ag(110), held at 230°C. The STM images reveal a self-assembled compact array of silicene nanoribbons with a very low density of defects. The NRs are tens of nanometers in length with a uniform width of 1.6 nm, and commensurate with the Ag(110) surface giving a (2x5) periodicity in the LEED measurements at saturation (1ML). We propose an atomic model of silicene nanoribbon with arm-chair edges. Careful scrutiny of the atomically resolved STM images, taken at several imaging conditions, indicates that not all silicon atoms in the NRs are observed. This is explained by a substantial buckling of the NRs and an electronic contribution due to the quantum confinement of the electrons in the NRs.

Acknowledgements
R.T. and H.O. acknowledge the financial support from the European Community FP7-ITN Marie-Curie Programme (LASSIE project, Grant Agreement No.238258) and the help of Professor J. L. Lemaire at the Observatoire de Paris. AK and MA would like to thank ISMO and STARM for support.

References

[1] S. Z. Butler, S. M. Hollen, L. Cao, Y. Cui, J. A. Gupta, H. R. Gutirrez, T. F. Heinz, S. S. Hong, J. Huang, A. F. Ismach, E. Johnston-Halperin, M. Kuno, V. V. Plashnitsa, R. D. Robinson, R. S. Ruoff, S. Salahuddin, J. Shan, L. Shi, M. G. Spencer, M. Terrones, W. Windl, and J. E. Goldberger, ACS Nano 7, 2898 (2013)
[2] Yang H, Heo J, Park S, Song H J, Seo D H, Byun K E, Kim P, Yoo I, Chung H J and Kim K 2012 *Science* **336** 1140
[3] Kara A, Enriquez H, Seitsonen A P, Lew Yan Voon L C, Vizzini S, Aufray B, and Oughaddou H 2012 *Surface Science Reports* **67** 1
[4] Cahangirov S, Topsakal M, Aktürk E, Sahin H and Ciraci S 2009 *Phys. Rev. Lett.* **102** 236804
[5] Teo B K and Sun X H 2007 *Chem. Rev.* **107** 1454
[6] Fagan S B, Baierle R J, Mota R, da Silva Z J R and Fazzio A 2000 *Phys. Rev.* B **61** 9994
[7] Lebègue S and Eriksson O 2009 *Phys. Rev.* B **79** 115409
[8] Guzman-Verri G G and Lew Yan Voon L C 2007 *Phys. Rev.* B **76** 75131
[9] Oughaddou H, Aufray B, Bibrian J P and Hoarau J Y 1999 *Surf. Sci* **429** 320
[10] Oughaddou H, Sawaya S, Goniakowski J, Aufray B, Le Lay G, Gay J M, Tréglia G, Bibérian J P, Barrett N, Guillot C, Mayne A J and Dujardin G 2000 *Phys. Rev. B* **62** 16653
[11] Oughaddou H, Léandri C, Aufray B, Girardeaux C, Bernardini J, Le Lay G, Bibérian J P and Barrett N 2003 *Appl. Surf. Sci* **1-5** 9781
[12] Léandri C, Oughaddou H, Gay J M, Aufray B, Le Lay G, Bibérian J P, Rangüis A, Bunk O and Johnson R L 2004 *Surf. Sci. Letter* **573** L369
[13] Aufray B, Kara A, Vizzini S, Oughaddou H, Landri C, Ealet B and Le Lay G 2010 *App. Phys. Lett.* **96** 183102
[14] Enriquez H, Vizzini S, Kara A, Lalmi B and Oughaddou H 2012 *J. Phys.Condens. Matter* **24** 314211
[15] Lalmi B, Oughaddou H, Enriquez H, Kara A, Vizzini S, Ealet B and Aufray B 2010 *Appl. Phys. Lett.* **97** 223109
[16] Tchalala M R, Enriquez H, Mayne A J, Kara A, RothS, Silly M.G, Bendounan A, Sirroti F, Greber T, Aufray B, Dujardin G, Ait Ali M and Oughaddou H 2013 *Appl. Phys. Lett.* **102** 083107
[17] Meng L, Wang Y, Zhang L, Du S, Wu R, Li L, Zhang Y, Li G, Zhou H, Hofer W.A and Gao HJ 2013 *Nanolett* **13** 685
[18] Lin C, Arafune R, Kawahara K, Tsukahara N, Minamitani E, Kim Y, Takagi N and Kawai M 2012 *App. Phys. Exp.* **5** 045802
[19] Feng B, Ding Z, Meng S, Yao Y, He X, Cheng P, Chen L and Wu K 2012 *Nano Lett.* **12** 3507
[20] Jamgotchian H, Colignon Y, Hamzaoui N, Ealet B, Hoarau J Y, Aufray B and Bibérian J P 2012 *J. Phys. Cond. Mat.* **24** 172001
[21] Majzik Z, Tchalala M R, Ŝvec M, Hapala P, Enriquez H, Kara A, Mayne A J, Dujardin G, Jelínek P and Oughaddou H 2013 *J. Phys. Condens. Matter* **25** 225301
[22] De Crescenzi M, Castrucci P, Scarselli M, Diociaiuti M, Chaudhari P.S, Balasubramanian C, Bhave T M and Bhoraskar S V 2005 *Appl. Phys. Lett.* **86** 231901
[23] Tchalala M.R, Ait Ali M, Enriquez H, Kara A, Lachgar A, Yagoubi S, Foy E, Vega E, Bendounan A, Silly M, Sirotti F, Nitshe S, Chaudanson D, Jamgotchian H, Aufray B, Mayne A.J, Dujardin G and Oughaddou H 2013 *J. Phys. Condens. Matter* **25** 442001
[24] Yang H, Mayne A J, Boucherit M, Comtet G, Dujardin G and Kuk Y 2010 *Nano Lett.***10** 943
[25] Cai J, Ruffieux P, Jaafar R, Bieri M, Braun T, Blankenberg S, Muoth M, Seitsonen A P, Saleh M, Feng X, Mllen K and Fasel R 2010 *Nature* **466** 470
[26] van der Lit J, Boneschanscher M P, Vanmaekelbergh D, Ijs M, Uppstu A, Ervasti M, Harju A, Liljeroth P and Swart I 2013 *Nat. Comms.* **4** 2023
[27] Owen J H G, Miki K, Koh H, Yeom H W and Bowler D R 2002 *Phys. Rev. Lett.* **88** 226104
[28] Soukiassian P, Semond F, Mayne A and Dujardin G 1997 *Phys. Rev. Lett.* **79** 2498
[29] De Padova P, Léandri C, Vizzini S, Quaresima C, Perfetti P, Olivieri B, Oughaddou H, Aufray B and Le LayG 2008 *Nano Lett.* **8** 2299
[30] Kara A, Vizzini S, Léandri C, Ealet B, Oughaddou H, Aufray B and LeLay G 2010 *J. Phys.Condens. Matter* **22** 045004
[31] Bena C and Simon L 2011 *Phys. Rev.* B **83** 115404

Two dimensional Si layer epitaxied on $LaAlO_3(111)$ substrate: RHEED and XPS investigations

C Ben Azzouz[1], A Akremi[1], M. Derivaz [2], J L Bischoff [2], M Zanouni[2,3], and Didier Dentel [2,a]

[1] Université de Carthage, Laboratoire de physique des matériaux, Unité de service commun spectromètre de Surfaces, Faculté des Sciences de Bizerte, 7021 Jarzouna Bizerte, Tunisia
[2] Institut de Science des Matériaux de Mulhouse (IS2M), UMR 7361 CNRS-UHA, Université de Haute Alsace, 3bis rue Alfred Werner, 68093 Mulhouse, France.
[3] Equipe de Recherche en Mécanique, Matériaux et Métallurgie, Université Abdelmalek Essaâdi, FST, ancienne route de l'aéroport, km 10, ziaten BP 416, 90050 Tangier-Morocco

E-mail: [a]`didier.dentel@uha.fr`

Abstract. We report the epitaxial growth of one Silicon monolayer on the $LaAlO_3(111)$ substrate, a high-κ crystalline oxide. Structural and chemical properties were investigated in-situ using reflection high energy electron diffraction (RHEED) and X-ray photoelectron spectroscopy (XPS). The deposition was achieved by molecular beam epitaxy in the temperature range RT-$500°C$. A two-dimensional epitaxial growth mode is observed for a deposition at temperature between $300°C$ and $500°C$. The deposited single layer is formed by two dimensional (2D) structures of Si.

1. Introduction

Dimensionality is one of the most important parameters which can exhibit dramatically different properties, depending on its crystal structure arrangement in a 0, 1, 2 or 3 dimensional (D) [1]. Currently there is an increasing interest in two-dimensional conducting materials such as graphene, a two-dimensional honeycomb lattice of carbon atoms [2]. Since the discovery of graphene and the important advancements in this research field [3], strong efforts have been investigated on graphene like materials (theoretically and experimentally), especially on silicon [4]. Silicene, the considered equivalent of graphene for silicon [5] forming a new synthetic atomic-thin two dimensional hexagonal silicon allotrope, has recently generated a very strong interest. It was first synthesized by silicon deposition on silver (110) and (111) [6, 7, 8, 9, 10]. While silicon is right below carbon in the periodic table of elements, valence orbitals do not sp^2 hybridize as easily as for its smaller counterpart [11]. Also it is difficult to form a freestanding 2D layer. Nevertheless, recently a theoretical study predict a possible approach to grow these graphite-like nanolattice by using a self-assembled layer-by layer growth method [12]. Furthermore Fleurence et al. demonstrated the successful growth of epitaxial silicene on a diboride surface [11].

Our group began studying silicon adsorption on lanthanum aluminate $LaAlO_3$ (LAO) surfaces a high-κ crystalline oxide several years ago. LAO is a highly attractive material due to its promising properties, including high value of dielectric constant (24), good thermodynamic stability with respect to Si [13, 14] and large band offsets (>2 eV) with Si [15, 16]. Indeed the growth of organized layer of silicon is expected, because on one hand the lattice mismatch between silicon and LAO(100) is of only about 1% when the Si lattice is rotated by 45° around an axis perpendicular to the surface with respect to the LAO lattice, on the other hand no interdiffusion was observed between silicon and LAO [17].

In this paper we have studied the growth of one silicon monolayer deposited by molecular beam epitaxy (MBE) on a LAO(111)-$(2\sqrt{3} \times 2\sqrt{3})R30°$ reconstructed surface, at different substrate temperatures. In the ($300°C$, $500°C$) temperature range a two dimensional epitaxial silicon structure has been successfully obtained on the LAO(111) surface.

2. Experiments

The experiments were performed in an a custom-built ultra-high vacuum (UHV) system with a base pressure of about $P = 10^{-10} mbar$, equipped with in-situ crystallographic and chemical characterization techniques such as reflection high-energy electron diffraction (RHEED) and X-ray photoelectron spectroscopy (XPS). LAO(111) substrates were purchased from CrysTec GmbH. After an ex-situ treatment in organic solvent, the LAO(111) substrate was cleaned in-situ with a heating during one hour at $900°C$ in order to obtain the $(2\sqrt{3} \times 2\sqrt{3})R30°$ reconstruction, the signature of a clean surface. The substrate temperature was controlled with an optical pyrometer. During the annealing, the pressure was maintained lower than $P = 10^{-9} mbar$. To evaporate Si, an electron gun was used. The deposition rate is about 0.25 monolayer/min and was controlled with a quartz microbalance. One monolayer (ML) is defined as the ideal atomic density in a Si (111) plane, i.e. $P = 7.8 \cdot 10^{14} at./cm^2$. The thicknesses indicated in this article were determined from the growth time assuming constant deposition rate. Both LAO(111) surface and Si growth were monitored in real time with a Staib Instrument RHEED system operating at an accelerating voltage of 25kV. XPS spectra were carried out using an hemispherical photoelectron spectrometer (Omicron EA 125) and a conventional AlKα (hν =1486.6 eV) photon source (Omicron DAR 400). The electron acceptance angle is about ±8° and the analysis area (diameter) does not exceed 1.5 mm. XPS chemical species core levels are analysed with pass energy of 20 eV. An overall resolution of ~1 eV for Ag$3d_{5/2}$ at 885.5 eV was used as an external reference. Binding energy shifts due to charge effects, was compensated by setting La$4d_{3/2}$ peak at 106.6 eV. In order to have a surface sensitive analysis geometry, the polar angle between the spectrometer and the normal of the surface is fixed at 50°. Core level spectra were fitted using CasaXPS software version 2.3.15 with symmetric mixed Gaussian-Lorentzian components.

3. Results and discussion

Figure 1 shows the RHEED evolution along the $< 11\bar{2} >$ (figures 1(a)-(c)) and the $< 1\bar{1}0 >$ (figures 1(d)-(f)) azimuths, of 1 ML Si deposit as a function of deposition temperature The presence of Kikuchi lines on the RHEED patterns reveal the excellent flatness of the surface. Starting from a streaky diagram corresponding to a clean and flat LAO(111) surface (figures 1(a),(d)), we observe an intensity attenuation of integral and fractional streaks after the deposit of 1 ML Si at $300°C$ (figures 1(b),(e)) and $500°C$ (figures 1(c),(f)), but the integral streaks are still visible on the RHEED patterns along $< 1\bar{1}0 >$ and $< 11\bar{2} >$ azimuths. The persistence of the integral streaks related to LAO(111) surface asserts that the growth is epitaxial and 2D. The system does not follow the thermodynamic predictions which consist in an immediate 3D growth mode. The single layer deposit on LAO(111) maintained at high temperatures ($300°C$ and $500°C$) during the growth is a two dimensional (2D) structure of Si. It seems that the reconstruction will govern the orientation and the anchorage of the Si deposit because the $(2\sqrt{3} \times 2\sqrt{3})R30°$ diffraction signature disappears immediately after the deposit begin. Nevertheless, as the theoretical model of the $(2\sqrt{3} \times 2\sqrt{3})R30°$ surface reconstruction is never precisely depicted in the literature, it is not possible to propose an atomic model of the relationship between the 2D Si layer and the LAO(111) surface.

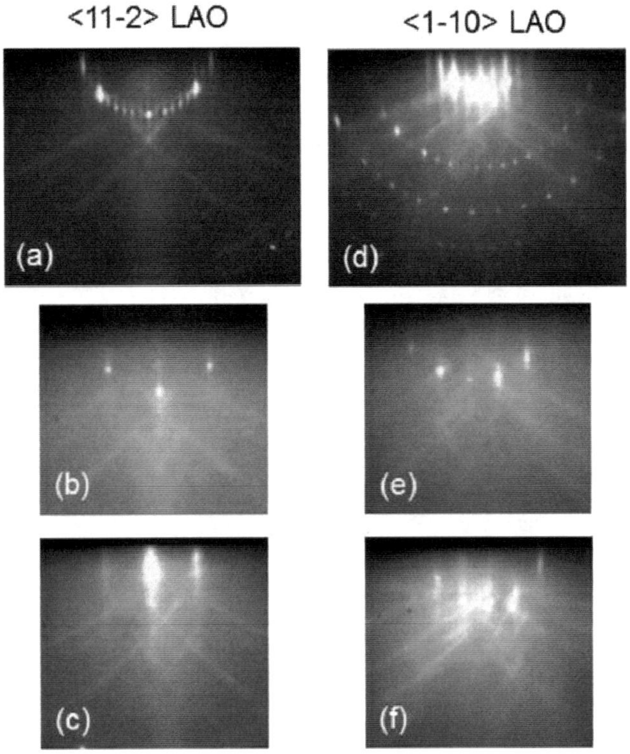

Figure 1. RHEED pattern evolution, along the $< 11\bar{2} >$ azimuth (a-c) and $< 1\bar{1}0 >$ azimuth (d-f) for the clean $LaAlO_3$ substrate ((a),(d)), after deposit at $300°C$ ((b),(e)) and $500°C$ ((c),(f)) of 1 ML Si.

In order to test the structural stability of the layer, the Si layer deposited at $500°C$ has been annealed at $600°C$ during 45 min (figure 2). The RHEED diagrams recorded after this thermal treatment along the $< 11\bar{2} >$ azimuth and the $< 1\bar{1}0 >$ azimuth are reported in figure 2(a) and 2(b) respectively. No drastic change has been pointed out on the diffraction patterns demonstrating a good thermal stability of the deposited 2D Si layer in spite of the thermodynamic predictions. It seems that the 2D Si layer is a new equilibrium shape on the LAO(111) surface.

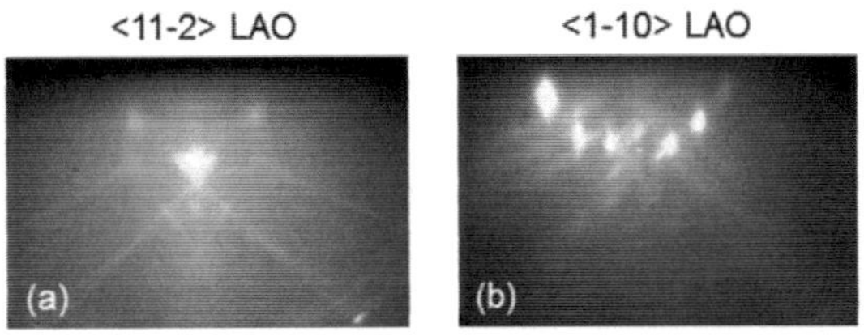

Figure 2. RHEED patterns after annealing at $600°C$ of the Si monolayer deposit at $500°C$, along the $< 11\bar{2} >$ (a) and $< 1\bar{1}0 >$ (b) azimuths.

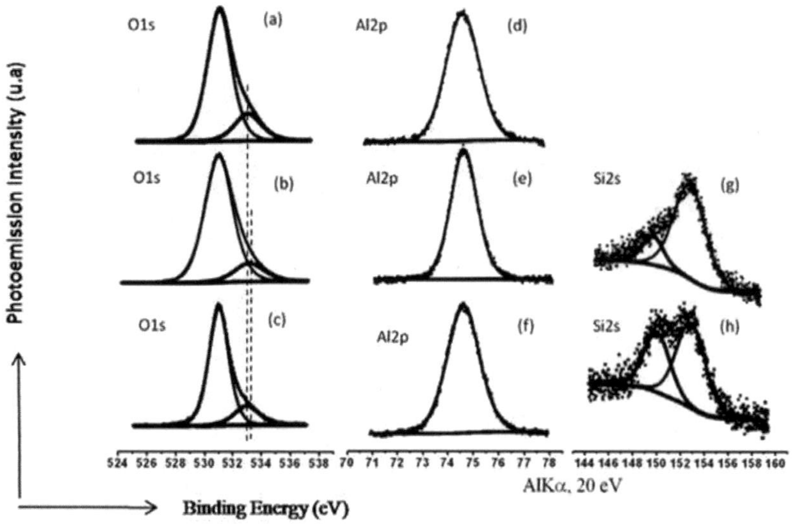

Figure 3. XPS O1s (a-c), Al2p (d-f) and Si2s (g,h) core level peak evolution as a function of the deposition temperature of 1ML of Si on $LaAlO_3(111)$: for the clean substrate (a,d), at $300°C$ (b,e,g) and at $500°C$ (c,f,h).

In order to determine the chemical composition of the deposited silicon monolayer as well as the Si/LAO interface quality, XPS analyses have been performed. The Si2s core level has been chosen because the Si2p component is overlapped with La$4d_{5/2}$ one. Figure 3 displays the evolution of O1s, Al2p and Si2s core level spectra, recorded in the case of clean LAO substrate (a),(d) and after the deposition of 1 silicon ML at $300°C$ (b, e and g) and at $500°C$ (c, f and h). First a complete C desorption is confirmed on the survey spectra (not shown here) after the LAO cleaning process. The O1s peak represented in figure 3(a) is recorded on a clean substrate and presents two components separated of 2 eV. The most intense component located at 531.3eV corresponds to the O environment in the LAO substrate. The second one, which appears at 533.3 eV, seems to be a surface component indicating that O atoms are probably involved in the LAO(111)-$(2\sqrt{3} \times 2\sqrt{3})R30°$ surface reconstruction. Its energy position could be attributed to Al-O environment indicating that LAO(111) surface could be Al terminated. Nevertheless, the Al$2p$ core level presented on figure 3(d) exhibits only one component corresponding to the Al environment in the LAO substrate. The Al oxidation on top of the LAO does not lead to detectable chemical shift (0.2 eV) with respect to our experimental resolution (1 eV) and probably appears in order to complete the AlO_2 octahedral blocks (AlO_{6-}) constituting the LAO substrate. Except for an intensity decrease of Al2p spectra, correlated to the amount of Si, neither the form nor the full with at half maximum (FWHM) are affected by the deposit (Figure 3(d)-(f)). These observations are extensible to the La$3d_{3/2}$ peaks (not shown here) indicating that no covalent bonds are formed during the Si deposition with these elements. Whatever the temperature used in this work, Si2s peak deconvolution reveals two components (Figure 3(g),(h)). While the less intense component, located at 150.5 eV, is attributed to $Si - Si_4$ environments of pure Si [18, 19], the second one is shifted on 2.7 eV on the high binding energy side leading to a $Si - SiO_3$ and/or $Si - Si_3O$ environments [20] indicating the formation of Si-O covalent bonds. This bond formation is ascertained on the $O1s$ spectra by the appearance of a new component located around 533 eV, characteristic of $Si - O_x$ (Figure 3(b),(c)), which replaces the surface component observed after cleaning and discussed before (Figure 3(a)). The position and the intensity of this component are highly connected to the x value explaining the slight difference observed as function of the deposit temperature.

The Si*2s* spectra of a Si single monolayer on a flat LAO(111) surface at $300°C$ (b,e,g) and at $500°C$ exhibit two components which have been previously discussed. The intensity of the $Si - Si_4$ component increases with the temperature. These observations strengthen our hypothesis of a progressive formation of a Si crystalline 2D-plane on top of the LAO(111) substrate which cover a more important part of the surface at $500°C$ than at $300°C$. At the same time, the intensity of the component related to the Si-O bonds decreases. The bonds between the Si layer and the LAO substrate, already observed on the O1s spectra (Figure 3(b),(c)), are confirmed. But at $500°C$ it seems that a larger part of the Si atoms are linked to other Si atoms and less to oxygen atoms as for $300°C$. The Si plane is progressively formed by increasing the deposition temperature. Nevertheless, the component located on the high binding energy side is most intense than the component associated to the pure Si, ascertaining the presence of Si-O covalent bounds in order to maintain a 2D Si layer on top of the LAO(111) surface.

4. Conclusion

We have carried out a structural and chemical growth study of 1 ML of silicon on $LaAlO_3$ (111) surface. We have shown that depending on the LAO(111) substrate temperature during silicon growth, a two dimensional epitaxial structure is observed. The annealing at 600C of the single silicon monolayer deposit at $500°C$ does not promote the formation of 3D structures ascertaining the stability of the two dimensional deposit.

5. Acknowledgement

This work was partially supported by the Hubert Curien Partnerships from the French ministry of foreign affair under the Volubilis project $n°$ MA/11/249.

References

[1] Kroto H W, Fischer J E and Cox D E 1993 *The Fullerenes* (Oxford, Pergamon)
[2] Zberecki K, Wierzbicki M, Barnas J and Swirkowicz R 2013 *Physical Review* B **88** 115404
[3] Geim A K 2009 *Science* **324** 1530
[4] Vogt P, De Padova P, Quaresima C, Avila J, Frantzeskakis E, Asensio M C, Resta A, Ealet B and Le Lay G 2012 *Phys. Rev. Lett.* **108** 155501
[5] Novoselov K S, Geim A K, Morozov S V, Jiang D, Zhang Y, Dubonos S V, Grigorieva I V and Firsov A A 2004 *Science* **306** 666
[6] Kara A , Leandri C, Davila M E, De Padova P, Ealet B, Oughaddou H, Aufray B and Le Lay G 2009 *J. Supercond. Nov. Magn.* **22** 259
[7] Le Lay G Aufray B, Leandri C, Oughaddou H, Biberian J P, De Padova P, Davila M E, Ealet B and Kara A 2009 *Appl. Surf. Sci.* **256** 524
[8] Aufray B, Kara A, Vizzini S, Oughaddou H, Leandri C, Ealet B and Le Lay G 2010 *Appl. Phys. Lett.* **96** 183102
[9] De Padova P, Quaresima C, Ottaviani C, Sheverdyaeva P M, Moras P, Carbone C, Topwal D, Olivieri B, Kara A, Oughaddou H, Aufray B and Le Lay G 2010 *Appl. Phys. Lett.* **96** 261905
[10] De Padova P, Kubo O, Olivieri B, Quaresima C, Nakayama T, Aono M and Le Lay G 2012 *Nano Lett.* **12** 5500
[11] Fleurence A, Friedlein R, OzakiT, Kawai H, Wang Y and Yamada-Takamura Y 2012 *Phys. Rev. Lett.* **108** 245501
[12] Houssa M, Pourtois G, Alfanas'ev V V and Stesmans A 2010 *Appl. Phys. Lett.* **97** 112106
[13] Lu X B, Zhang X, Huang R, Lu H B, Chen Z H and Xiang W F 2004 *Appl.Phys. Lett.* **84** 2620
[14] Goel N, Majhi P, Tsai W, Warusawithana M, Schlom D G and Santos M B 2007 *Appl. Phys. Lett.* **91** 093509
[15] Devos I and Boulenc P. 2007 *Appl. Phys. Lett.* **90** 072906
[16] Mi Y Y, Yu Z, Wang S J, Lim P C, Foo Y L and Huan A C H 2007 *Appl. Phys Lett.* **90** 181925
[17] Mortada H, Dentel D, Derivaz M, Bischoff J L, Denys E, Moubah R, Uhlaq-Bouillet C and Werckmann J 2011 *J. Crys. Growth* **323** 247

[18] Miller M L and Linton R W 1985 *Analysis Chemistry* **57** 2314
[19] Zhdan P A, Shepelin A P, Osipova Z G and Sokolovskii V D 1979 *J. Catalysis* **58** 8
[20] Himpsel F J, Mc Feely F R, Taleb-Ibrahimi A, Yarmoff J A and Hollinger G 1988 *Phys. Rev. B* **38** 6084

Atomic structure of the $(2\sqrt{3} \times 2\sqrt{3})R30°$ of silicene on Ag(111) surface

Hanna Enriquez[1,a], **Abdelkader Kara**[2,b], **Andrew J. Mayne** [1,c], **Gérald Dujardin**[1,d], **Haik Jamgotchian**[3,e], **Bernard Aufray**[3,f] **and Hamid Oughaddou**[1,4,g]

[1] Institut des Sciences Moléculaires d'Orsay, ISMO-CNRS, Bâtiment 210, Université Paris-Sud, 91405 Orsay, France
[2] Department of Physics, University of Central Florida, Orlando, Florida, 32816 USA
[3] CINaM-UPR, 7325, CNRS, Campus de Luminy, 13288 Marseille Cedex 09,France
[4] Département de Physique, Bâtiment Neuville II, 5 Mail Gay-Lussac, 95031 Neuville sur Oise, Cergy-Pontoise Cedex, France

E-mail: [a]hanna.enriquez@u-psud.fr, [b]Abdelkader.Kara@ucf.edu, [c]andrew.mayne@u-psud.fr, [d]gerald.dujardin@u-psud.fr, [e]haik@cinam.univ-mrs.fr, [f]aufray@cinam.univ-mrs.fr, [g]hamid.oughaddou@u-cergy.fr,

Abstract. The deposition of one monolayer of silicon on a Ag(111) substrate induces the formation of silicene structures exhibiting different ordered phases, including a $(2\sqrt{3}\times2\sqrt{3})R30°$, a (4×4) and a $(\sqrt{13}\times\sqrt{13})R13.9°$ superstructures. In this paper we focus on the $(2\sqrt{3}\times\sqrt{3})R30°$ phase. Using a combination of scanning tunneling microscopy and LEED observations, we show that this phase corresponds to a 11° rotation of the silicene sheet relative to the substrate orientation.

1. Introduction

Silicene is considered to be a promising novel material for nano-electronics as it will naturally benefits from the vast Si-based Research and Development knowledge (e.g. band gap opening, doping, making electrical contacts, etc)[1, 2]. It has been shown that it has an intrinsic stability and presents electronic properties almost identical to those of graphene [3, 4]. According to theoretical calculations, silicene presents a slight buckling of 0.44 Å between the two inequivalent sublattices, and an in-plane lattice parameter of 3.83 Å [3, 4]. On the other hand, the synthesis of freestanding silicene sheets has progressed only to the stage of silicon nanotubes [5] and multilayer silicene using chemical methods [6]. Thus, growing silicene on solid substrates in particular silver, gold and iridium remains the primary method [7, 8, 9, 10, 11].

Experimentally, the synthesis of silicene has been achieved on the Ag(110) surface where parallel silicene nano-ribbons (NRs) exactly 16 Å wide were obtained [8]. These nanoribbons are potentially very interesting because they have a perfectly well defined edge structure and show 1D dispersion [8] with particular quantum interference effects similar to graphene [12]. To achieve similar results, other 1D structures such as graphene nanoribbons must be synthesized chemically [13], Bi nanolines on Si(001) require a higher formation temperature ($600°C$) [14], and Si nanowires on SiC did not show any 1D conductivity [15].

Now on the Ag(111) surface, a continuous two dimensional (2D) sheet of silicene can be grown. The first experimental study [9] observed the honeycomb silicene structure with a well-defined $(2\sqrt{3} \times 2\sqrt{3})R30°$ superstructure. Soon after this breakthrough, several groups reported the successful growth of silicene on Ag(111) in the $(2\sqrt{3} \times 2\sqrt{3})R30°$ phase, as well as other ordered phases with (4×4) and $(\sqrt{13} \times \sqrt{13})R13.9°$ superstructures [16, 17, 18]. These ordered phases have been interpreted as corresponding to different orientations of the silicene sheet relative to the Ag(111) surface [18, 19].

The $(2\sqrt{3} \times 2\sqrt{3})R30°$ structure was obtained upon deposition of one monolayer of silicon on a Ag(111) substrate kept at a temperature of about $230°C$ [9]. A long-range ordered structure was observed as in both LEED and STM. The STM images show a honeycomb arrangement at the atomic scale, which was interpreted as a local atomic-resolution of a silicene sheet. However, these results were questioned recently due to the small Si-Si first neighbor interatomic distances of 1.9 Å deduced from STM topography [20]. The claim that the STM images do not correspond to a silicene layer but to a bare Ag(111) surface relies on the postulate that the "honeycomb" structure observed is due to a contrast inversion, in other words to a graphically inverted image or to a tip effect [20].

In this paper we present conclusions from a very careful analysis of the STM images on wich accurate drift corrections have been performed. We have determined that the "honeycomb structure" is actually rotated by about 11° with respect to the orientation of the lattice of the bare substrate. In a two-dimensional model of a silicene sheet deposited on a (111) silver layer, we find that this rotation angle of 11° leads to a $(2\sqrt{3} \times 2\sqrt{3})R30°$ superstructure of the substrate, in agreement with the LEED pattern.

2. Experiments

The experiments were performed under ultra-high vacuum (UHV) using the standard tools for surface preparation and characterization: an ion gun for surface cleaning, a low energy electron diffractometer (LEED) for structural characterization, a STM for surface characterization at the atomic scale working at room temperature (RT), and an Auger electron spectrometer (AES) for chemical surface analysis and calibration of the silicon coverage. The Ag(111) sample was cleaned by several cycles of sputtering ($600eV Ar^+$ ions, $P = 5 \cdot 10^{-5} mbar$) and annealing at $450°C$ until a sharp p(1x1) LEED pattern was obtained. Silicon was evaporated by direct current heating of a piece of Si wafer, and the one monolayer coverage was calibrated using AES. During silicon deposition, the Ag(111) surface was held at a temperature of $(200 - 230)°C$, the temperature being controlled by a thermocouple located close to the sample.

3. Results and discussion

Figure 1 shows the $(2\sqrt{3} \times 2\sqrt{3})R30°$ superstructure observed by LEED after the deposition of one monolayer of silicon on the Ag(111) substrate kept at the above mentioned temperature $(200 - 230°C)$. The unit cell and the spots of the (1x1) are highlighted in figure 1. It is clear from the LEED that a $(2\sqrt{3} \times 2\sqrt{3})R30°$ superstructure has been formed upon deposition of a one monolayer of silicon.

After taking the LEED measurements, the sample was transferred in vacuum for STM measurements. It is important to stress that one needs to perform measurements under the same conditions as the LEED. Because silicene presents many phases, this important step is crucial for assigning the electronic properties to a given phase. Figures 2 (a) and (b) show the STM images obtained respectively on the bare Ag(111) substrate and after the deposition of one silicon ML, that gave the LEED pattern in figure 1. Given there is no rotation of the sample before and after, it is clear that the silicene sheet is rotated by 11° with respect to the bare Ag(111) surface as the pairs of red, green and blue lines show. The STM images were recorded at high current (3 nA) and low bias voltage (-10 mV). The atomic resolved STM images reveal a

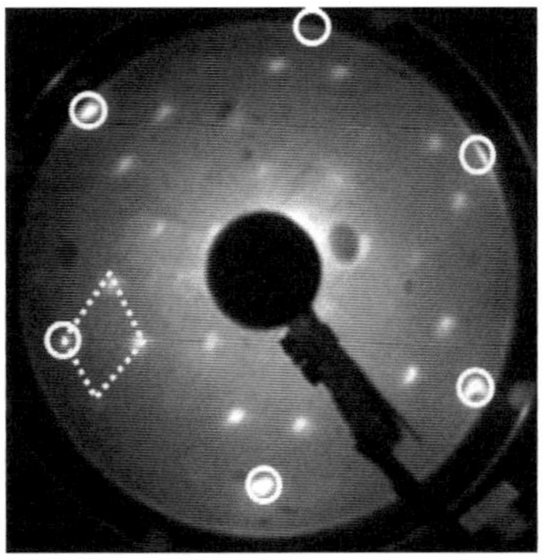

Figure 1. $(2\sqrt{3} \times 2\sqrt{3})R30°$ LEED pattern corresponding to the deposition of 1 silicon monolayer. The unit cell (dotted diamond) and the the spots (small circles) of the (1x1) are highlighted.

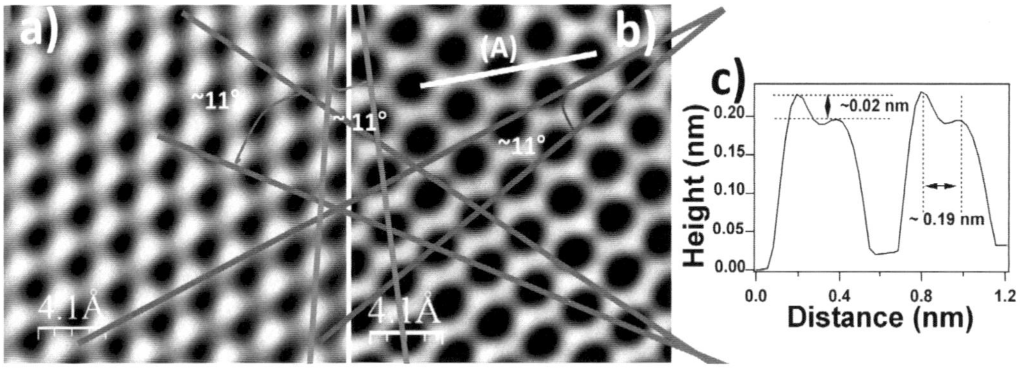

Figure 2. $(2\sqrt{3} \times 2\sqrt{3})R30°$ LEED pattern corresponding to deposition of 1 silicon monolayer. a) A filled-state atomically resolved STM image of the clean Ag(111) surface. b) A filled-state atomically resolved STM image of the same sample after deposition of one silicon monolayer ($U_t = -10mV$, $I_t = 3nA$). The pairs of red, blue and green lines show the rotation of the silicene sheet with respect to the clean Ag(111) c) Line-profile joining neighboring Si atoms along the direction (A) indicated in (b).

highly ordered honeycomb lattice silicon structure (Figure 2(b)), while the line profile along line (A) shows a corrugation of 2 Å (from peak to peak) and a height difference between neighboring Si atoms (buckling) of 0.2 Å (Figure 2(c)).

Based on the STM images of the bare Ag(111) (Figure 2(a)), the silicene sheet (Figure 2(b)) and the LEED pattern (Figure 1), we propose the two-dimensional model shown in Figure 3. In this model, a $(\sqrt{7} \times \sqrt{7})R19.1°$ supercell of the silicene sheet coincides with a $(2\sqrt{3} \times 2\sqrt{3})R30°$ supercell of the silver layer.

A 19.1° rotation of the basal chains of the honeycomb silicene sheet gives an effective angle of $(30 - 19.1)° = 10.9°$ with respect to the axes of the Ag(111, which matches very well our data (Figure 2).

On the basis of the above observations, we can definitely rule out the conclusions of Le Lay et al. claiming that our STM image corresponds to the bare silver surface based on a graphical inversion of the image or due to a tip effect [20]. Indeed, in the case of an inversion of contrast, the angle between the observed honeycomb structure and that of the bare Ag(111) would be 0° as shown in Figure 4. Moreover, the model proposed here and in our previous study [9] is supported by two other recent studies [17, 18].

As we have shown, our careful analysis of the alignment of the observed honeycombs against

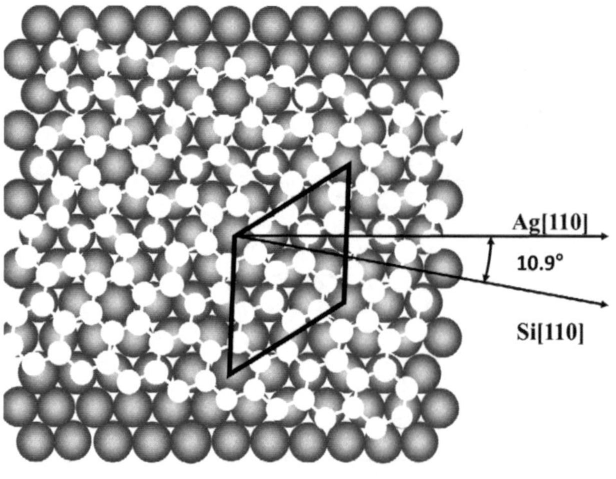

Figure 3. A ball model of silicene on Ag(111) derived from both STM images (Fig 2(a),(b)) and from the observed $(2\sqrt{3} \times 2\sqrt{3})R30°$ LEED pattern shown in Fig 1(a). Silicon and silver atoms are colored in white and grey respectively. Silver atoms showing the $(2\sqrt{3} \times 2\sqrt{3})R30°$ superstructure are colored in green.

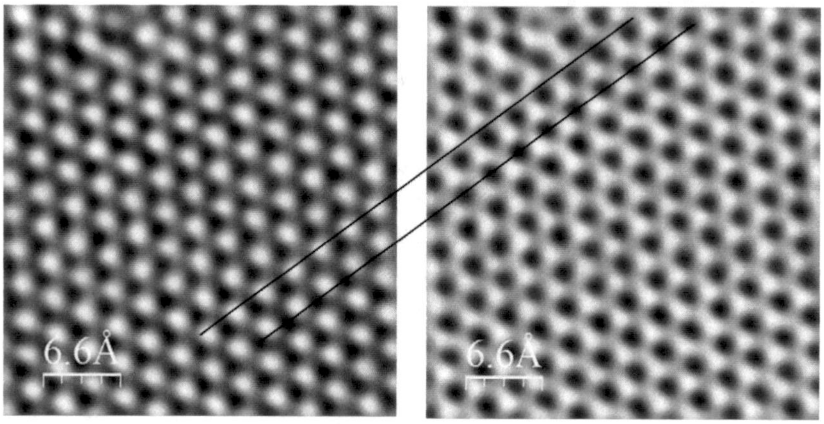

Figure 4. (a) STM image recorded on a clean bare Ag(111)-1x1. (b) graphically inverted image of (a). The axis of the honeycomb-like arrangement resulting from the inversion of the contrast is aligned with the high symmetry axis of figure (a) as highlighted by the parallel dark lines.

the high symmetry direction of the substrate clearly shows an 11° rotation (Figure 2), in agreement with the proposed model (Figure 3). This is irrefutable evidence that STM image in Fig. 2(a) corresponds to a silicene sheet on Ag(111).

Now we will discuss the interatomic distances. The line profile (fig 2c) shows a lateral Si-Si neighbor distance of (1.9 ± 0.1) Å, and not 1.7 Å as has been sometimes stated [16, 20]. This distance is within the range of the values expected for a silicene sheet on Ag(111) [19]. From an experimental point of view, the distance between two protrusions derived from a STM image is not trivial for two reasons. The first is that a projected distance onto the surface plane is a lower bound for the actual interatomic distance. A strong buckling gives a projected distance that is lower than the actual Si-Si distance. The second is a consequence of the STM measurement;

the surface density of states dominates the STM image over the geometrical contribution. In extreme cases, a maximum in the density of states may not correspond to the position of an atom [21, 22]. However, in a recent experiment combining both STM and non-contact atomic force microscopy (AFM) on a silicene sheet in the (4×4) phase, the results clearly show the coincidence between the protrusions in the STM images and the positions of the silicon atoms [23].

Since the maxima in the STM images correspond to the density of states of the Si atoms at the tip position, we can deduce that the actual atomic buckling is more than the observed buckling of 0.2 Å (fig 2c), as is observed on the clean Si(100) 2x1 surface [24]. As a consequence, the apparent lateral Si-Si distance of (1.9 ± 0.1) Å is shorter than the real Si-Si bond length [9, 18, 19]. DFT calculations confirmed that the silicene sheet corresponding to a $(2\sqrt{3}\times2\sqrt{3})R30°$ superstructure is strongly buckled, with an out-of-plane buckling of 1Å, silicon interatomic distances varying from 2.28 Å to 2.37 Å (giving a projected lateral distance of 2 Å). Due to this strong buckling only two silicon atoms per supercell were imaged in STM topography with a bias voltage of 1V [17, 18] whereas the whole honeycomb atomic structure is resolved in the present study at a bias of 10 mV. This bias dependent behavior has been observed on the clean Si(100) 2x1 surface [24] and also on graphene [25] and is determined by the coupling of the surface states to the bulk [26].

4. Conclusion

In conclusion, we evidence that the honeycomb structure of the $(2\sqrt{3}\times2\sqrt{3})R30°$ superstructure obtained upon deposition of Si/Ag(111) is due, neither to the inversion of the contrast of the STM image, nor to the contribution of the second silver layer. The orientation of the silicene superstructure deviates by about 11° from that of the substrate. The silicon interatomic distance deduced from STM images is always a lower bound of the actual distance due to the substantial buckling of the adsorbed silicene, in contrast with the flat graphene. The model proposed is in agreement our data and those of other groups [17, 18].

References

[1] Teo B K and Sun X H 2007 *Chem. Rev.* **107** 1454
[2] Dong L, Bush J, Chirayos V, Solanki R, Jao J, Ono Y, Conley J.F and Ulrich B.D 2005 *Nano Letters* **5** 2112
[3] Lebègue S and Eriksson O 2009 *Phys. Rev.* B **79** 115409
[4] Cahangirov S, Topsakal M, Aktürk E, Sahin H and Ciraci S 2009 *Phys. Rev. Lett.* **102** 236804
[5] De Crescenzi M, Castrucci P, Scarselli M, Diociaiuti M, Chaudhari P.S, Balasubramanian C, Bhave T M and Bhoraskar S V 2005 *Appl. Phys. Lett.* **86** 231901
[6] Tchalala M.R, Ait Ali M, Enriquez H, Kara A, Lachgar A, Yagoubi S, Foy E, Vega E, Bendounan A, Silly M, Sirotti F, Nitshe S, Chaudanson D, Jamgotchian H, Aufray B, Mayne A.J, Dujardin G and Oughaddou H, *J. Phys. Condens. Matter* accepted
[7] Kara A, Enriquez H, Seitsonen A P, Lew Yan Voon L C, Vizzini S, Aufray B, and Oughaddou H 2012 *Surface Science Reports* **67** 1
[8] Aufray B, Kara A, Vizzini S, Oughaddou H, Landri C, Ealet B and Le Lay G 2010 *App. Phys. Lett.* **96** 183102
[9] Lalmi B, Oughaddou H, Enriquez H, Kara A, Vizzini S, Ealet B and Aufray B 2010 *Appl. Phys. Lett.* **97** 223109
[10] Tchalala M R, Enriquez H, Mayne A J, Kara A, RothS, Silly M.G, Bendounan A, Sirroti F, Greber T, Aufray B, Dujardin G, Ait Ali M and Oughaddou H 2013 *Appl. Phys. Lett.* **102** 083107
[11] Meng L, Wang Y, Zhang L, Du S, Wu R, Li L, Zhang Y, Li G, Zhou H, Hofer W.A and Gao HJ 2013 *Nanolett* **13** 685
[12] Yang H, Mayne A.J, Boucherit M, Comtet G, Dujardin G and Kuk Y 2010 *Nano Lett.* **10** 943
[13] Van der Lit J, Boneschanscher M P, Vanmaekelbergh D, Ijas M, Uppstu A, Ervasti M, Harju A, Liljeroth P and Swart I 2013 *Nat. Comm.* **4** 2023
[14] Owen J H G, Miki K, Koh H, Yeom H W and Bowler D.R 2002 *Phys. Rev. Lett.* **88** 226104
[15] Soukiassian P, Semond F, Mayne A and Dujardin G 1997 *Phys. Rev. Lett.* **79** 2498
[16] Lin C, Arafune R, Kawahara K, Tsukahara N, Minamitani E, Kim Y, Takagi N and Kawai M 2012 *App. Phys. Exp.* **5** 045802

Journal of Physics: Conference Series **491** (2014) 012004 doi:10.1088/1742-6596/491/1/012004

[17] Feng B, Ding Z, Meng S, Yao Y, He X, Cheng P, Chen L and Wu K 2012 *Nano Lett.* **12** 3507
[18] Jamgotchian H, Colignon Y, Hamzaoui N, Ealet B, Hoarau J Y, Aufray B and Bibérian J P 2012 *J. Phys. Cond. Mat.* **24** 172001
[19] Enriquez H, Vizzini S, Kara A, Lalmi B and Oughaddou H 2012 *J. Phys.Condens. Matter* **24** 314211
[20] Le Lay G, De Padova P, Resta A, Bruhn T and Vogt P 2012 *J. Phys. D* **45** 392001
[21] Knall J, Pethica J B, Todd J D and Wilson J H 1991 *Phys. Rev. Lett.* **66** 1733
[22] Altibelli A, Joachim C and Sautet P 1996 *Surf. Sci.* **367** 209
[23] Majzik Z, Tchalala M R, Ŝvec M, Hapala P, Enriquez H, Kara A, Mayne A J, Dujardin G, Jelínek P and Oughaddou H 2013 *J. Phys. Condens. Matter* **25** 225301
[24] Hata K, Yasuda S and Shigekawa H 1999 *Phys. Rev.* B **60** 8164
[25] Yang H, Cejas C, Mayne A.J, Dujardin G and Kuk Y 2013 *App. Phys. Lett.* **102** 223104
[26] Dujardin G, Mayne A J and Rose F 2002 *Phys. Rev. Lett.* **89** 036802

Journal of Physics: Conference Series **491** (2014) 012005 doi:10.1088/1742-6596/491/1/012005

Tailoring of the structural, energetic and electronic properties of silicene-based nanostructures

Xianqing Lin[1], **Chao Lian**[1] **and Jun Ni**[1*]

Department of Physics and State Key Laboratory of Low-Dimensional Quantum Physics, Tsinghua University, Beijing 100084, P. R. China

E-mail: *junni@mail.tsinghua.edu.cn

Abstract. Silicene-based nanostructures are highly promising for future applications in electronics as silicon is an important element for conventional semiconductor industry. In our recent works, we have studied the effect of external electric filed, strain and metal adatoms on tailoring the structural, energetic and electronic properties of silicene-based nanostructures using first-principles and tight-binding methods. We find that half-metallicity can be realized in zigzag silicene nanoribbons when applying electric field across the nanoribbon width. Under small tensile strain, some armchair silicene nanoribbons are predicted to have linear energy dispersion around the Fermi level. And strain-induced structural phase transitions are observed in silicene bilayers in which strong covalent interlayer bonds form. When metal atoms are adsorbed on silicene, several metal adatoms obtain a larger binding energy than the cohesive energy of the bulk metal and the bonding between the metal atoms and silicene can be ionic or covalent.

1. Introduction

Silicene, the silicon counterpart of graphene, possesses many intriguing properties similar to graphene[1, 2, 3, 4]. The honeycomb π-orbital network makes silicene show linear energy dispersion near the Fermi level (E_F)[1, 2]. Thus the quasiparticles near E_F in silicene show Dirac-type behavior of motion. Since silicon cannot form layed graphite-like material, much efforts have been devoted to grow silicene on the substrates[5, 6, 7, 8, 9, 10, 11]. Recently, silicene has been successfully fabricated on the Ag (111) surfaces[10, 11]. Similar to graphene, Si atoms in silicene form two sublattices. However, the two sublattices in silicene are not in the same plane with a buckled height between the two sublattices of about 0.46 Å[1]. In addition to the structural difference, the Si atoms in silicene have lower electronegativity than C in graphene. Then some silicene-based nanostructures can acquire rather distinct properties from graphene nanostructures. And we have been focusing on exploiting the rich electronic and magnetic properties of silicene-based nanostructures[12, 13]. Besides the pristine silicene nanostructures, the effects of external fields such as electric field[13] and strain[14, 15] as well as functionalization with metal adatoms[16] on tailoring the the structural, energetic and electronic properties of silicene nanostructures have been studied using first-principles and tight-binding methods.

Among the silicene-based nanostructures, silicon nanotubes and silicene nanoribbons are the one-dimensional (1D) structures corresponding to silicene. In experiements, equally spaced silicene nanoribbons have been observed on the Ag (110) surfaces[5, 8]. And silicon nanotubes with a hollow structure have also been reported to be grown from silicon monoxide[17]. While silicene and graphene are both semimetalic, the 1D structures can be semiconductors as a result of the confinement effect. In both theoretical and experimental studies, graphene nanoribbons are demonstrated to be semiconducting[18, 19]. However, whether a graphene nanotube are metal or semiconductor depends on the chirality for tubes with a large diameter[20]. For silicon nanotubes, our first-principles calculations show that the surfaces of nanotubes are buckled with Si atoms having different distances to the axis of the tube[12]. And this difference in structures makes silicon nanotubes have distinct electronic properties from CNT. In contrast to CNT, the silicon nanotubes with a large diameter are all semiconductors[12]. The external electric field, strain and metal adatoms provide ways of tailoring various properties of silicene structures. For graphene nanoribbons with zigzag edges (ZGNR), it is demonstrated that the transverse electric field can make the ZGNR undergo a transition from semiconductor to half-metal[21]. Half-metallicity can also be realized in zigzag silicene nanoribbons (ZSiNR) under transverse electric field. Since silicene and silicene nanoribbons are obtained through growth on the substrate, the strain may play an important role in the structural and electronic properties of these nanostructure[14]. In addition to the monolayer silicene, silicene bilayers have also been observed in experiments[11]. The structures of silicene bilayers under strain have been studied as the silicene bilayers grown on the substrate tend to be strained compared to the pristine structure[15]. Adsorption of metal atoms is an effective way of functionalizing silicene and graphene nanostructures. Nanostructures adsorbed with metal atoms may be applied for superconductivity[22], hydrogen storage[23, 24, 25, 26], catalyst[27, 28], as well as tuning the electronic and magnetic properties[29]. However, the binding energies of all studied metal adatoms on graphene are smaller than the cohesive energies of the bulk metals[29]. As a result, metal adatoms on graphene with high coverage may tend to form clusters[30]. Silicene has larger hexagons than graphene and the silicene sheet is buckled while the graphene sheet is planar. Moreover, partial sp^3 characteristics are present in the Si-Si σ bonds in silicene rather than complete sp^2 hybridization in graphene. Then silicene with some metal adatoms can exhibit unique structural, energetic, electronic properties compared with graphene[16].

The outline of this paper is as follows: Section II presents the effect of transverse electric field on the electronic structures of ZSiNR. In Sec. III, we show the structural and electronic properties of strained armchair silicene nanoribbons (ASiNR) and silicene bilayers. In Sec. IV, the functionalization of silicene with metal adatoms is discussed. Section V is the summary.

2. Zigzag silicene nanoribbons under transverse electric field

Silicene nanoribbons can be considered to be cut along a specific direction from the silicene sheet. Silicene nanoribbons are also buckled. And the silicene nanoribbons are passivated by hydrogen atoms. For N_Z-ZSiNR, N_Z is used to denote the number of zigzag lines across the ribbon width. Using first-principles calculations, the stable magnetic state of ZSiNR and the behavior of ZSiNR under transverse electric field have been investigated[13].

A ZSiNR has an antiferromagnetic (AFM) ground state and is semiconducting[13]. The stabilities of the nonmagnetic (NM) state, ferromagnetic (FM) state, and AFM state of ZSiNR are calculated. It is found that the AFM state has the lowest energy. For example, the energy of the AFM state is 25 meV/unit less than the FM state and the energy of the FM state is 20 meV/unit less than the NM state for 6-ZSiNR. For the antiferromagnetic ZSiNR, the two edges have opposite spin orientations. For 6-ZSiNR, the spin-up and spin-down bands are degenerate with a band gap of 0.17 eV. When applying electric field across the width of the nanoribbon, ZSiNR undergoes a transition from semiconductor to half-metal with the increasing electric field[13]. As an example, for 6-ZSiNR, the band gap of up-spin remain larger than 0.17 eV for electric field in the range of $0 \sim 0.17$ V/Å. While for the down-spin, the band gap decreases with the increase of E and become closed when E reaches 0.15[13]. From the calculations, it is

Journal of Physics: Conference Series **491** (2014) 012005 doi:10.1088/1742-6596/491/1/012005

found that the critical electric field E_c satisfies a universal behavior as wE_c=2.6 V, where w is the width of the ZSiNRs. The value of wE_c is smaller than that of ZGNRs[21], which suggests that it can be easier for a ZSiNR to become half-metals than the ZGNR having the same width as the ZSiNR.

3. Strained armchair silicene nanoribbons and silicene bilayers

3.1. Band gaps of ASiNR tubed by strain

The Dirac type dispersion in silicene and silicon nanoribbons indicates a high mobility of electrons near the Fermi level. It is demonstrated that the silicon nanoribbons with armchair edges (ASiNR) are semiconductors with small gaps[1, 31, 13], recognized as a suitable material. Our first-principles calculations have shown that the structural parameters of 8-ASiNR agree well with the experimental observation [32]. The successful experimental synthesis of the 8-ASiNRs makes this purpose practical. To realize the function of field effect devices, a variation of the conductivity of ASiNRs will be needed, which is traditionally tuned by a gate voltage. However, a quite large gate voltage should be applied to open a small gap in silicene.[33] Other methods such as hydrogenation will open a gap but meanwhile break the linear dispersion near the Fermi level[34]. It is important to find a method to tune the band gaps with the Dirac-type dispersion maintained.

We find that the strain is an effective method to tune the band gap of the SiNRs[14]. Moreover, the Dirac type dispersion is kept near the Fermi level during the modulation. As shown in Fig. 1(a), the linear dispersions still hold except for the Γ point when the strain is around the ε_0. The band gap of 11-ASiNR is 39 meV without strain, it becomes 58 meV when $\varepsilon = 3\%$. When applied a strain of $\varepsilon = 1.24\%$, the band gap becomes quite small. The charge densities of the highest occupied crystal orbital (HOCO) and the lowest unoccupied crystal orbital (LUCO) shows that the strain turn the HOCO state in Fig. 1(a) into LUCO state in Fig. 1(c), and meanwhile turn the HOCO state in Fig. 1(a) into HOCO state in Fig. 1(c). When applied a strain of $\varepsilon = 1.24\%$, the two types of charge density distributions mixed, indicates that the HOCO state and LUCO state are degenerate, and the band gap seem to vanish.

However, the band gaps from DFT calculations can not give an exact zero value because the numerical errors in the computations are inevitable. It is important to judge whether the band gap is exactly equal to zero, which indicates a semiconductor-semimetal-semiconductor phase transition. Therefore, we use the tight-binding model to prove the existence of a zero point in the band gap under certain strain[14]. For the (3p+2)-ASiNRs under different strains, the band gap occurs at k=0, drawn from the DFT calculations (Fig. 2(a)). The Hamiltonian of the ASiNRs with the the same Si-Si bond lengths is the unperturbed Hamiltonian H_0. The perturbation Hamiltonian is H' and thus the total Hamiltonian is $H = H_0 + H'$. The tight-binding model can be represented by a two-leg ladder lattice system, as shown in Fig. 2(b). The unperturbed Hamiltonian H_0 can be written as

$$H_0 = \begin{pmatrix} T & t_0 I \\ t_0 I & T \end{pmatrix}_{2N\times 2N} \tag{1}$$

where

$$T = \begin{pmatrix} 0 & t_0 & 0 & 0 & \cdots & 0 & 0 & 0 & 0 \\ t_0 & 0 & t_0 & 0 & \cdots & 0 & 0 & 0 & 0 \\ 0 & t_0 & 0 & t_0 & \cdots & 0 & 0 & 0 & 0 \\ \vdots & \vdots & \vdots & \vdots & \ddots & \vdots & \vdots & \vdots & \vdots \\ 0 & 0 & 0 & 0 & \cdots & t_0 & 0 & t_0 & 0 \\ 0 & 0 & 0 & 0 & \cdots & 0 & t_0 & 0 & t_0 \\ 0 & 0 & 0 & 0 & \cdots & 0 & 0 & t_0 & 0 \end{pmatrix}_{N\times N} \tag{2}$$

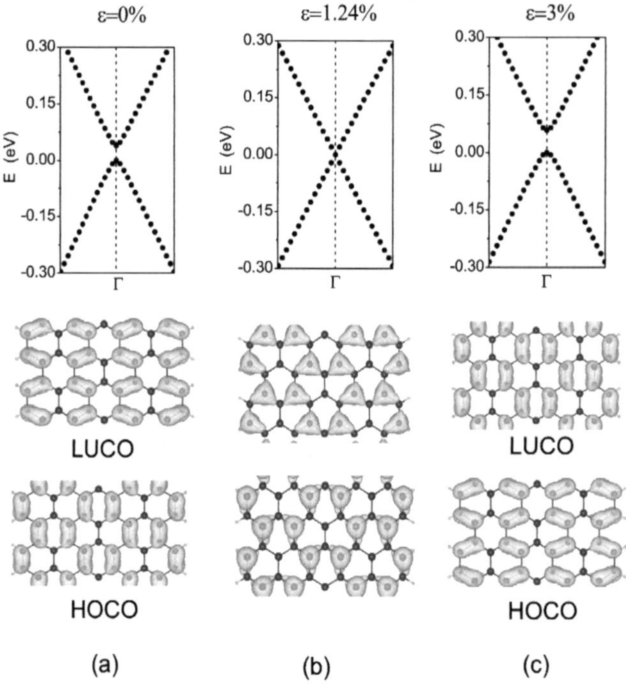

Figure 1. (Color online)(top) The detailed band structures of 11-ASiNRs near the Γ point with the strain of (a) $\varepsilon = 0\%$ (b) $\varepsilon = 1.24\%$ and (c)$\varepsilon = 3\%$. The highest occupied energies are set to zero.(middle)(bottom) The partial charge densities of the lowest unoccupied crystal orbital (LUCO) and the highest occupied crystal orbital (HOCO) for 11-ASiNR with the strain of (a)$\varepsilon = 0\%$, (b)$\varepsilon = 1.24\%$ and (c)$\varepsilon = 3\%$. The isosurface of partial charge densities is $0.005e/\AA^3$. With the strain of $\varepsilon = 1.24\%$, the LUCO and HOCO are degenerate.[14]

and I is the $N \times N$ unit matrix. t_0 is the hopping integrals between the nearest neighbors. When a strain is applied, the bond are stretched, leading to the variation of the hopping integrals. The perturbation Hamiltonian is

$$H' = \begin{pmatrix} A & B \\ B & A \end{pmatrix}_{2N\times 2N} \tag{3}$$

where

$$A = \begin{pmatrix} 0 & \delta t_2' & 0 & 0 & \cdots & 0 & 0 & 0 & 0 \\ \delta t_2' & 0 & \delta t_2 & 0 & \cdots & 0 & 0 & 0 & 0 \\ 0 & \delta t_2 & 0 & \delta t_2 & \cdots & 0 & 0 & 0 & 0 \\ \vdots & \vdots & \vdots & \vdots & \ddots & \vdots & \vdots & \vdots & \vdots \\ 0 & 0 & 0 & 0 & \cdots & \delta t_2 & 0 & \delta t_2 & 0 \\ 0 & 0 & 0 & 0 & \cdots & 0 & \delta t_2 & 0 & \delta t_2' \\ 0 & 0 & 0 & 0 & \cdots & 0 & 0 & \delta t_2' & 0 \end{pmatrix}_{N\times N} \quad (4)$$

$$B = \begin{pmatrix} \delta t_1 & & & & & \\ & \delta t_1 & & & & \\ & & \ddots & & & \\ & & & \delta t_1 & & \\ & & & & \delta t_1 & \\ & & & & & \delta t_1' \end{pmatrix}_{N\times N} . \quad (5)$$

Using perturbation theory, the band gap has been obtained:

$$E_g = |E_1' - E_2'| = | - 2\delta t_2 + 2\delta t_1 + \frac{2(\delta t_1' - \delta t_1) - 4(\delta t_2' - \delta t_2)}{p+1}|. \quad (6)$$

The hopping parameters is assumed to be scaled in Harrison form[31] and the bond lengths under various strains are linearly fitted near the strain $\epsilon = 0$. As shown in Fig. 2(c), we find the band gaps of 8-ASiNR, 11-ASiNR and 14-ASiNR vanish with small strains and the zero point right shift with the increase of N_a, the same as the results of the ab initio calculations(Fig. 2(a))[14].

3.2. Strain-induced Phase Transitions in silicene bilayers

Besides that the monolayer silicene have been grown experimentally[10, 11], it is also reported that a bilayer silicene phase coexist and with similar structural and electronic properties with the monolayers[11]. However, the first-principle calculations shows that the original low buckled silicene monolayer will become a complete sp^3 Si(111) sheet. With relaxation, this stacking silicene bilayer turn into the reconstructed silicene bilayer (RSBL), which is the global stable structure[36, 37]. This contradiction comes from the substrate effect in the experiments which is not taken into consideration in the calculations. Among these effects strain is an important factor. Thus, it is important to discuss the changes of the silicene bilayer structures under strain[15].

We study the reconstructed structures of RSBL under isotropic tensile strains[15], as shown in Fig. 3(a) and 3(b). As shown in Fig. 3(c), with the increasing strain, the buckle lengths decrease to 0.08Å under the strain of 10.8%. After that, the RSBL undergoes a phase transition, with decrease of 0.074eV/atom in the total energy. This phase (Fig. 3(b)) is similar as the AA stacking graphene bilayer and thus named as graphene-like silicene bilayer(GSBL). We find that this GSBL phase exists in a wide range of strain. Although the GSBL emerges under the strain over 10% from the barrierless phase transitions, the GSBL has already become the most stable structure when the strain reaches 6.5%, as shown in Fig. 3(d). As we gradually decrease the strain from 20% to 0% on the GSBL, the total energy decreases to the minimum of -4.96eV/atom at the strain of 8.3% and increases after that. Before the strain decreases to 6.5%, the GSBL is always the most stable structure. The SSBL and RSBL will eventually turn into the GSBL when the strain is larger than 6.5%, but with energy barriers. The extra strain is needed for the RSBL to overcome the barriers and the barrierless phase transitions occur at $\epsilon = 10.3\%$.

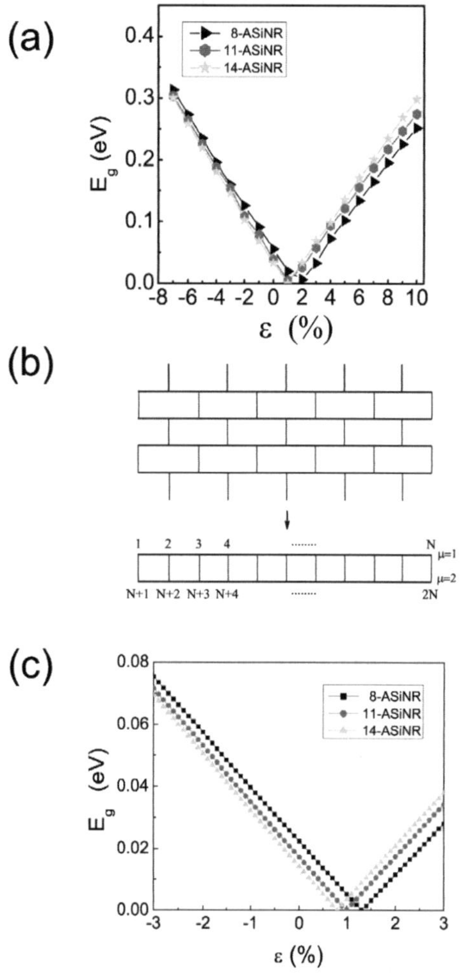

Figure 2. (Color online) (a) The band gaps E_g of (3p+2)-ASiNRs as a function of strain. (b)The two-leg ladder model of ASiNRs at Γ point. (c) The band gaps of (3p+2)-ASiNRs as a function of strain in tight-binding calculations.[14]

The silicon atoms in the GSBL are not three-coordinated as silicene, but four-coordinated like the sp^3 silicon materials. However, different from the common sp^3 silicon materials, the four nearest neighbors of the silicon atoms in the GSBL do not form a regular tetrahedron. It indicates that the hybridization in GSBL are novel and not similar with neither the sp^2 hybridization nor the sp^3 hybridization. To reveal the hybridization type of the GSBL, we carry on a tight-binding calculation[15].

The model is shown in Fig. 4(a) and the Hamiltonian can be expressed as:

$$H = \begin{pmatrix} H_{AA} & H_{AB} & H_{AC} & 0 \\ H_{BA} & H_{BB} & 0 & H_{BD} \\ H_{CA} & 0 & H_{CC} & H_{CD} \\ 0 & H_{DB} & H_{DC} & H_{DD} \end{pmatrix}, \tag{7}$$

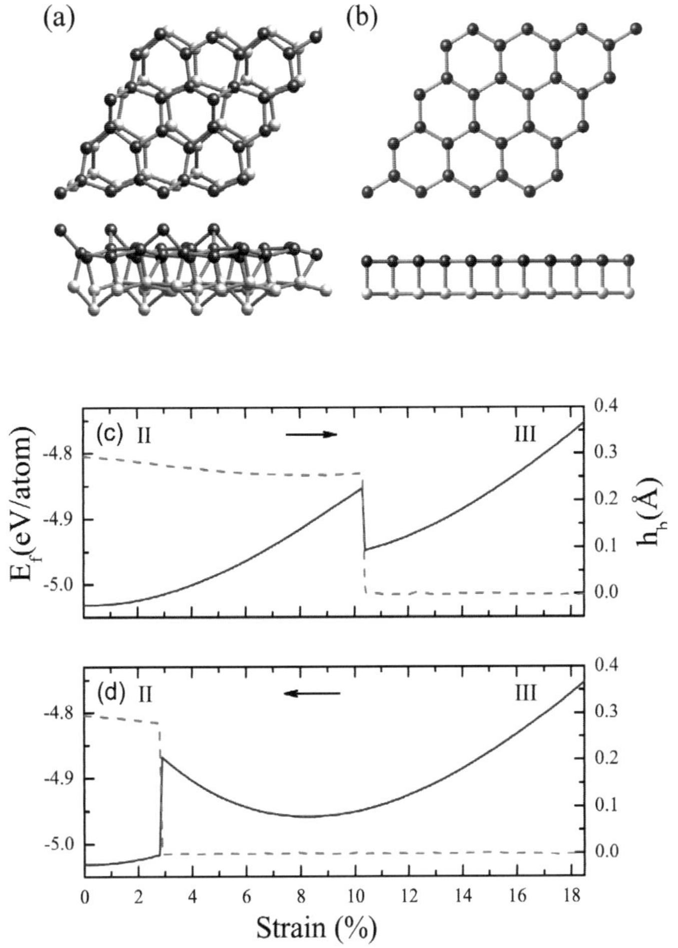

Figure 3. (Color online) The structures of the (a) RSBL and (b) GSBL. The energies (solid lines) and buckle lengthes h_b (dash lines) of (c)RSBL and (d)GSBL with strains. The arrows represent the variations of the strain.[15]

where $H_{i,j} = H^{\dagger}_{j,i}$ is a matrix describing the interaction between atom i and atom j. The symmetry demands $H_{AA} = H_{BB} = H_{CC} = H_{DD}$ and $H_{AB} = H_{CD}, H_{AC} = H_{BD}$. There are four atomic orbitals per silicon atom:s, p_x, p_y and p_z. Thus in the Slater-Koster scheme[20, 38], $H_{i,j}$ is a 4×4 matrix. H_{AA} and H_{AB} describe the intralayer interactions:

$$H_{AA} = \begin{pmatrix} \varepsilon_s & 0 & 0 & 0 \\ 0 & \varepsilon_p & 0 & 0 \\ 0 & 0 & \varepsilon_p & 0 \\ 0 & 0 & 0 & \varepsilon_p \end{pmatrix}, H_{AB} = \begin{pmatrix} h_{s^A s^B} & h_{s^A p_x^B} & h_{s^A p_y^B} & 0 \\ h_{p_x^A s^B} & h_{p_x^A p_x^B} & h_{p_x^A p_y^B} & 0 \\ h_{p_y^A s^B} & h_{p_y^A p_x^B} & h_{p_y^A p_x^B} & 0 \\ 0 & 0 & 0 & h_{p_z^A p_z^B} \end{pmatrix}. \quad (8)$$

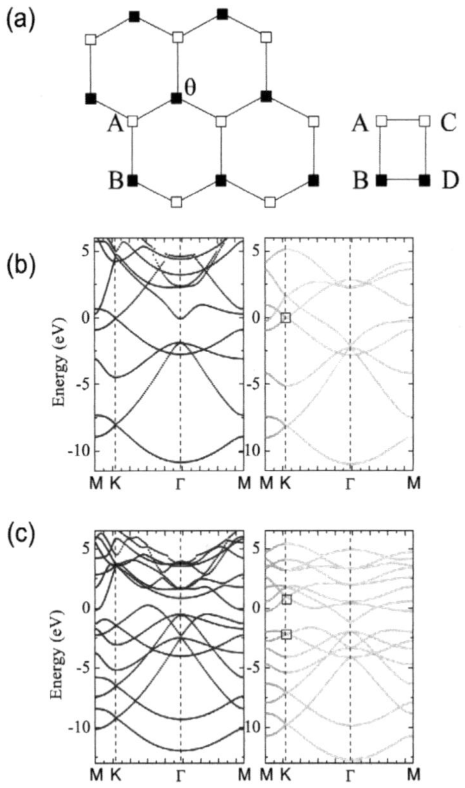

Figure 4. (Color online) (a) The schematic diagram of the tight-binding model. The left panel shows the top view and the right panel shows the side view. A B C and D represent different sublattices. θ is the direction angle of the vector between the atom A and atom B. (b) and (c) The band structures of the PSML and GSBL. The left panels and the right panels show the first-principles and tight-binding results, respectively.[15]

H_{AC} describes the interlayer interactions in GSBL:

$$H_{AC} = \begin{pmatrix} h_{s^A s^C} & 0 & 0 & h_{s^A p_z^C} \\ 0 & h_{p_x^A p_x^C} & 0 & 0 \\ 0 & 0 & h_{p_y^A p_y^C} & 0 \\ h_{p_z^A s^C} & 0 & 0 & h_{p_z^A p_z^C} \end{pmatrix}, \tag{9}$$

We find that the phase transitions occurs under certain tensile strains for SSBL and RSBL, both with no energy barrier[15]. After the transition, the SSBL and RSBL turn into a graphene-like silicene bilayer (GBSL) structure. Different from the AB stacking graphene bilayer with the Van der Waals interlayer interactions, the GBSL is combined in the AA stacking with the strong covalent bonds. Using the tight-binding model, we demonstrate that the hybridization of the electrons in the GSBL is the $sp^2 - sp^1$ hybridization without any sp^3 part.

We use the parameters determined from the first-principles energy band. No extra parameters are introduced when two planar silicene bilayer(PSML) form the GSBL. Therefore, the parameters are fitted with the first-principles band structure of the PSML, as shown in Fig. 4(b). With these parameters, as shown in Fig. 4(c), the band structure of GSBL is achieved. The tight-binding results agree well with the first-principles results, which means that the primary interactions have been included in our tight-binding model[15]. This is totally different from the weak Van der Waals interactions in the graphene bilayer, which can be viewed as a perturbation to the band structure of the graphene monolayer. It concludes that the intralayer sp^2 hybridization in the GSBL is the same as that in the PSML, since H_{AA} and H_{AB}, which describe the intralayer interaction, are exactly the same. In addition, because of the existence of the interlayer interaction H_{AC}, the properties of the GSBL are different from the two isolated PSMLs. We recall that the H_{AC} describe the hybridization just like the sp^1 hybridization in the 1D silicon chains. It demonstrates that the hybridization of the silicon atoms in GSBL is sp^2-sp^1 hybridization.

We tailor the GSBL into armchair nanoribbons (Fig. 5(a)) and zigzag nanoribbon (Fig. 5(b)). The band structure of bilayer zigzag silicon nanoribbon (BZSiNR) is similar to the monolayer ZSiNR (Fig. 5(c))[13]. While due to the sp^3 edges, the spin-correlation between two edges vanish and the ground state is nonmagnetic. The hydrogen atoms in the passivated edge lies out-plane, which indicates a sp^3 hybridization. As shown in Fig. 5(d) and 6(e), the narrower(N_a <11) bilayer armchair silicene nanoribbon (BASiNR) is semiconductor, the same as monolayer ASiNR, while the band gaps decrease with the widths increase. The periodicity of band gaps in monolayer ASiNRs is absent. A wider BASiNR(N_a ≥11)are metallic.

4. Silicene adsorbed with metal atoms

Adsorption of metal atoms is an effective way of functionalizing nanostructures. Much effort has been devoted to studying the behavior of metal adatoms on graphene[22, 29]. The metal atoms adsorbed on silicene can acquire rather distinct structural, energetic and electronic properties compared with those on graphene. A systematic investigation of the adsorption of 15 different metal atoms on silicene, including alkali and alkali-earth metals Li, Na, K, Ca, group III and IV metals Al, Ga, In, Sn and transition metals Ti, Fe, Co, Ni, Pd, Pt,Au, has been done[16].

Metal atoms have three possible adsorption sites, the hollow (H) site above the center of a Si hexagon, the top (TA) site above a Si atom belonging to sublattice A and the top (TB) site above a Si atom of sublattice B, as shown in Fig. 6(a). This is different from graphene, for which metal atoms can be adsorbed above the midpoint of a C-C bond (bridge site)[29]. Since graphene is planar, the TA and TB sites are equivalent. Li, Na, K, and Ca bind most strongly to the H site of silicene. Al, Ga, In, and Sn prefer the TA site, as shown in Fig. 6(b). For transition metal atoms, the most stable adsorption site is H except for Fe, whose favored site is TA. In the case of Fe at the H site, E_b is a little smaller than the TA site. The Fe adatom at the H site of silicene has a peculiar structure with the six Fe-Si bonds having almost equal length, which is about 2.30 $\AA$, close to the Si-Si bond length 2.39 $\AA$ in silicene, as shown in Fig. 6(c)[16].

Compared with graphene, metal atoms have much strong binding to silicene. The binding energy E_b is defined as $E_b(\mathrm{M}) = E(\mathrm{M}) + E(\mathrm{Silicene}) - E(\mathrm{M\text{-}Silicene})$, where E(M), E(Silicene), and E(M-Silicene) are the energies of an isolated metal atom, the silicene supercell and the adatom-silicene system. For silicene, Li, Na, K, Ca, Co, Ni, Pd, and Pt obtain a larger E_b than the cohesive energy of the bulk metal (E_c)[16]. However, all the 15 metal atoms considered have smaller E_b than E_c on graphene[29]. Therefore, metal adatoms on silicene are rather stable.

Silicene adsorbed with different metal atoms exhibit various electronic and magnetic

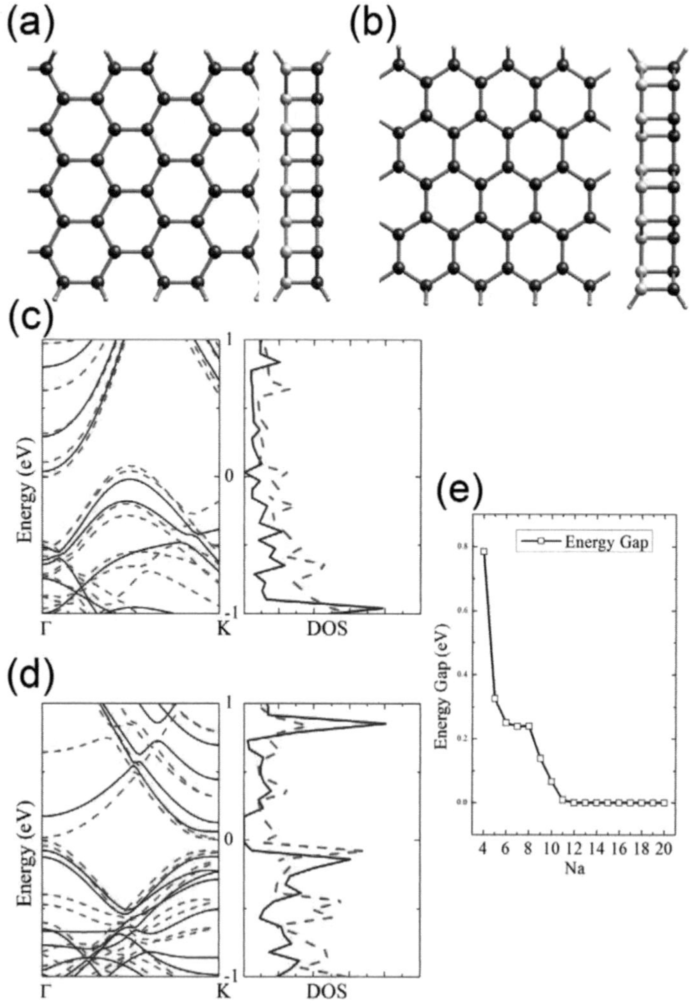

Figure 5. (Color online) (a)(b)The structures of 8-ABSiNR and 5-ZBSiNR. The left panel is the top view and the right panel is the side view. The black and grey balls represent the silicon atoms in different layers. (c) The energy band(left) and density of states(right) of 8-ABSiNR(solid lines) and 12-ABSiNR(dash lines). (d) The energy band(left) and density of states(right) of 5-ZBSiNR(solid lines) and 9-ZBSiNR(dash lines). (e) The energy gaps of ABSiNRs as a function of N_a.

properties[16]. When an alkali atom Li, Na or K is adsorbed on silicene, about one electron is transfered from the atom to silicene. E_F is shifted above the Dirac point (E_D) for about 0.39 eV, as shown in Fig. 7(b). The K state lies about 1.5 eV above E_F. Therefore, the bonding of alkali metal atoms to silicene is approximately ideal ionic. For the adsorption of Ca on silicene, the 3d states of Ca are hybridized with the silicene states around and above E_F, as shown in Fig. 8. The Ca adatom on silicene has a magnetic moment of 0.7 μ_B. The hybridization band near E_F of the up-spin is partially occupied, while the down-spin is totally empty. When the group III and IV metals Al, Sn, and transition metal atoms Ti, Fe, Co, and Au are adsorbed on silicene, hybridization between the adatom states and the silicene states occurs around the

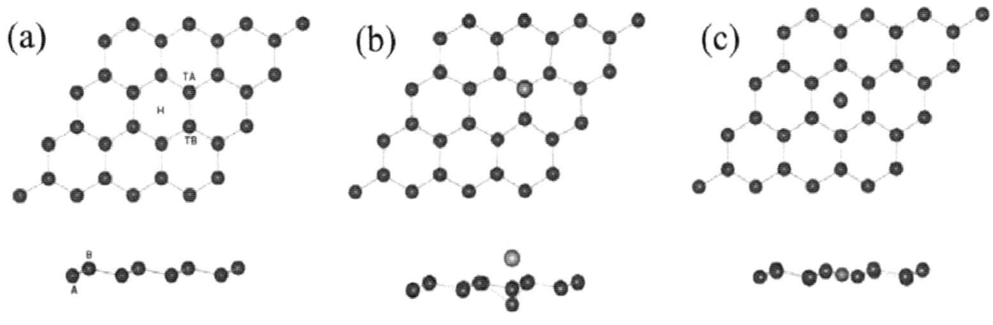

Figure 6. (color online) (a) The buckled hexagonal structure of silicene. The three possbile adsorption sites are labeded as H, TA, and TB. (b) and (c) The relaxed structure for Al@TA and Fe@H.[16]

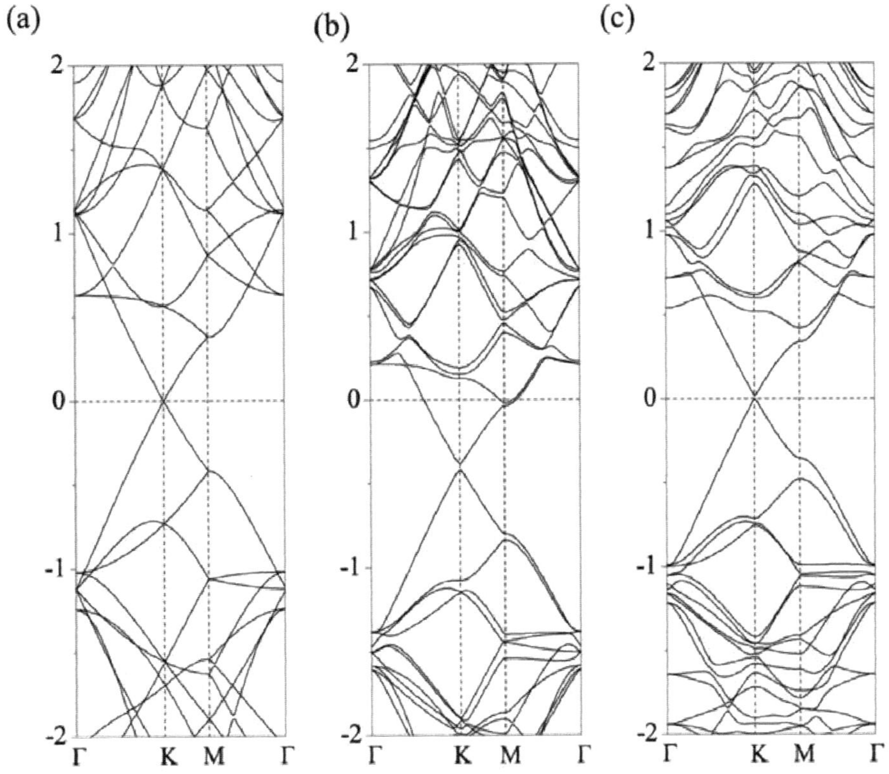

Figure 7. (color online) (a) Band structure of the 4 × 4 silicene supercell. (b) and (c) Band structures for the K and Pd adatoms on silicene, respectively.

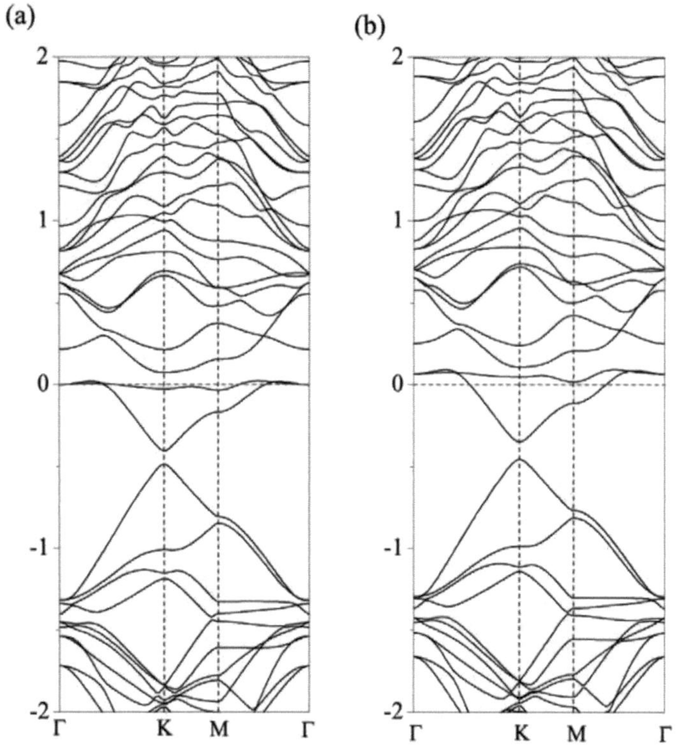

Figure 8. (color online) (a) Band structures of the spin-up states (a) and spin-down states (b) for the Ca adatom on silicene.

Dirac point. In contrast, for silicene with the Ni, Pd, and Pt adatoms, the hybridization states between the adatom d states and the silicene states lie rather far below the E_F with E_F located approximately at E_D. Figure 7(c) shows the band structure for Pd on silicene. The interaction between the silicene and Pd d states arise below the energy of -0.88 eV.

5. summary

In summary, we reviewed our recent works on the structural, energetic and electronic properties of silicene-based nanostructures modulated with external electric filed, strain and metal adatoms. We find that under external electric field across the width of the nanoribbon, the ZSiNR which has an antiferromagnetic ground state with the two edges having opposite spin orientations can become half-metalic. The ASiNR are semiconductors with the band gaps showing three series as a function with the width of the ribbon. And the band gap of ASiNR can be tuned by strain and the (3p+2)-ASiNRs are predicted to have linear energy dispersion under some small tensile strain. In addition, the isotropic tensile strain can make silicene bilayers undergo several barrierless structure phase transitions. For silicene bilayers, there exist strong covalent interlayer bonds. When metal atoms are adsorbed on silicene, metal adatoms can have different adsorption configuration on silicene from that on graphene and Li, Na, K, Ca, Co, Ni, Pd, and Pt obtain a larger binding energy than the cohesive energy of the bulk metal. The bonding between metal adatoms and silicene can be ionic or covalent. The alkali atoms bond ionically to silicene. There exits hybridization between the adatoms states and the silicene states for Ca, groups III and IV and transiton metal atoms adsorbed on silicene.

Acknowledgements
This research was supported by the National Natural Science Foundation of China under Grant nos. 11174171 and 10721404.

References

[1] Cahangirov S, Topsakal M, Aktürk E, Şahin H and Ciraci S 2009 *Phys. Rev. Lett.* **102** 236804
[2] Lebègue S and Eriksson O 2009 *Phys. Rev.* B **79** 115409
[3] Geim A K and Novoselov K S 2007 *Nature Mater.* **6** 183
[4] Castro Neto A H, Guinea F, Peres N M R, Novoselov K S and Geim A K 2009 *Rev. Mod. Phys.* **81** 109
[5] Leandri C, Lay G Le, Aufray B, Girardeaux C, Avila J, Dávila M E, Asensio M C, Ottaviani C and Cricenti A 2005 *Surf. Sci.* **574** L9
[6] Nakano H, Mitsuoka T, Harada M, Horibuchi K, Nozaki H, Takahashi N, Nonaka T, Seno Y and Nakamura H 2006 *Angew. Chem.* **118** 6451
[7] Sahaf H, Masson L, Léandri C, Aufray B, Lay G Le and Ronci F 2007 *Appl. Phys. Lett.* **90** 263110
[8] Padova P De, Quaresima C, Perfetti P, Olivieri B, Leandri C, Aufray B, Vizzini S and Lay G Le 2008 *Nano Lett.* **8** 271
[9] Kara A, Enriquez H, Seitsonen A P, Lew Yan Voon L C, Vizzini S, Aufray B and Oughaddou H 2012 *Surf. Sci. Rep.* **67** 1
[10] Vogt P, De Padova P, Quaresima C, Avila J, Frantzeskakis E, Asensio M C, Resta A, Ealet B and Lay G Le 2012 *Phys. Rev. Lett.* **108** 155501
[11] Chen L, Liu C, Feng B, He X, Cheng P, Ding Z, Meng S, Yao Y and Wu K 2012 *Phys. Rev. Lett.* **109** 56804
[12] Yang X and Ni J 2005 *Phys. Rev. B* **72** 195426
[13] Ding Y and Ni J 2009 *Appl. Phys. Lett.* **95** 083115
[14] Lian C, Yang Z and Ni J, *Chem. Phys. Lett.* **77** 561
[15] Lian C and Ni J 2013 *AIP Advances* **3** 052102
[16] Lin X and Ni J 2012 *Phys. Rev. B* **86** 075440
[17] Chen Y W, Tang Y H, Pei L Z and Guo C 2005 *Adv. Mater. (Weinheim, Ger.)* **17** 564
[18] Son Y-W, Cohen M L and Louie S G 2006 *Phys. Rev. Lett.* **97** 216803
[19] Han M Y, Ö zyilmaz B, Zhang Y and Kim P 2007 *Phys. Rev. Lett.* **98** 206805
[20] Saito R, Dresselhaus G and Dresselhaus M S 2003 Physical Properties of Carbon Nanotubes Imperial College Press, London
[21] Son Y-W, Cohen M L and Louie S G 2006 *Nature (London)* **444** 347
[22] Profeta G, Calandra M and Mauri F 2012 *Nature Phys.* **8** 131
[23] I. Cabria, M. J. López, and J. A. Alonso, J. Chem. Phys. **123**, 204721 (2005).
[24] Ataca C, Aktürk E, Ciraci Sand Ustunel H 2008 *Appl. Phys. Lett.* **93** 043123
[25] Ataca C, Aktürk E and Ciraci S 2009 *Phys. Rev.* B **79** 041406(R)
[26] Ao Z M and Peeters F M 2010 *Phys. Rev.* B **81** 205406
[27] Lu Y-H, Zhou M, Zhang C and Feng Y-P 2009 *J. Phys. Chem.* C **113** 20156
[28] Li Y, Zhou Z, Yu G, Chen W and Chen Z 2010 *J. Phys. Chem.* C **114** 6250
[29] Chan K T, Neaton J B and Cohen M L 2008 *Phys. Rev.* B **77** 235430
[30] Cabria I, López M J and Alonso J A 2010 *Phys. Rev.* B **81** 035403
[31] Sahin H, Cahangirov S, Topsakal M, Bekaroglu E, Akturk E, Senger R and Ciraci S 2009 *Phys. Rev.* B **80** 155453
[32] Lian C and Ni J 2012 *Physica* B **407** 4695
[33] Ni Z, Liu Q, Tang K, Zheng J, Zhou J, Qin R, Gao Z, Yu D and Lu J 2011 *Nano. Lett.* **12** 113
[34] Zhang P, Li X, Hu C, Wu S and Zhu Z 2012 *Phys. Lett.* A **376** 1230
[35] Kara A, Leandri C, Davila M, De Padova P, Ealet B, Oughaddou H, Aufray B and Lay G Le 2009 *J. Supercond. Novel Magn.* **22** 259
[36] Morishita T, Spencer M, Russo S, Snook I and Mikami M 2011 *Chem. Phys. Lett.* **506** 221
[37] Spencer M, Morishita T and Snook I 2012 *Nanoscale* **4** 2906
[38] Slater J and Koster G 1954 *Phys. Rev.* **94** 1498

Interdistance Effects on Flat and Buckled Silicene Like-bilayers

S. Naji[1,2], B. Khalil[1] ,H. Labrim[3], M.Bhihi[1], A. Belhaj[4], A. Benyoussef[1,5,6], M. Lakhal[1], A. El Kenz[1]

[1] LMPHE, (URAC 12), Faculté des Sciences, Université Mohammed V-Agdal, Rabat, Morocco
[2] Department of Physics, Faculty of Science,Ibb University, Ibb, Yemen
[3] Centre National de l'Energie, des Sciences et des Techniques Nucléaires, Rabat, Morocco
[4] Département de Physique, Faculté Polydisciplinaire, Université Sultan Moulay Slimane, Béni Mellal, Morocco
[5] Institute of Nanomaterials and Nanotechnology, MAScIR, Rabat, Morocco
[6] Hassan II Academy of Science and Technology, Rabat, Morocco

Abstract. Using ab intio numerical calculations based on the all-electron full-potential local-orbital minimum-basis scheme FPLO9.00-34, we discuss the interdistance effect on the energy gap of two parallel layers of the silicone systems. The like- bilayer systems we dealt with here are relying on a dynamic monolayer of silicene located at distance d along the normal direction z forming with a static one a (AA) stacking arrangement. In particular, we investigate the effect of the dynamic layer by varying the distance d starting from a distance around the bond length of Van der Waals. More precisely, we consider the flat and two buckled geometries in (AA) arrangements. The flat geometry is associated with the usual (AA) configuration appearing in the pure graphene material. For buckled geometry, we can distinguish two configurations. The first one corresponds to the usual buckled configuration that keeps the same vertical distance between the two layers atoms while the remaining one is obtained by reversing one silicene layer. We show that the band gap can be opened by simply varying the distance, starting around a Van der Waals distance, between two parallel silicene for flat and buckled geometries due to an electronic transition of electrons living in pz orbital states. Furthermore, we study the stability between the buckled and the flat configuration in the mono and bilayer system.

1. Introduction

The study of two dimensional materials (2DM) with honeycomb structure has been considered as a subject of great interest in lower energy physical models [1-10]. It becomes the most promised research topics in the condensed matter physics and related areas including high energy physics. A special interest is to investigate the electronic structure of some particular materials composed of atoms appearing in the fourth group of Mendeleev chemistry table [11,12]. It has been shown that honeycomb-structured models play a crucial role placed in nano-science and they could be considered as the most stable material systems used in the semiconductor electronic applications [5,13,14].

In the recent years, many efforts have been devoted to study the physics of such materials using different methods with appropriate approximations. Besides that, some theoretical and analytic developments have been also given supporting the obtained experimental results found in 2DM with success in isolating single layer material. This includes the discovery of the graphene from bulk natural graphite by Noveselov et al [1]. The very nice feature of this material is that near the so-called Dirac points the valence and the conduction bands intersect, exhibiting linear energy dispersion [15]. This research area has received an intense attention because the graphene behaves like a semiconductor with zero energy gap. Analytically, the corresponding low energy excitations are described by the Dirac equation dealing with 2 dimensional massless fermions [16].

On the other hand, there has been also a great interest in studying the silicene since it involves properties similar to those of the graphene [5,6,10]. For this reason, the silicene, which is a monolayer of the silicon atoms forming a 2D honeycomb lattice, can be also viewed as an alternative potential candidate for nanotechnology applications. This material has been synthesized and modeled using different approaches [17-21].

The results of single layer studies of 2DM with honeycomb structure have been extended to multi-layers systems. Special emphasis is put on bilayer like systems relying on inter-interactions between two parallel layers [22]. In such systems, the electronic and transport properties can be modified radically according to layer stacking arrangements [23]. It turns out that there are various types of layer arrangements where the most symmetric configurations are the direct (AA) and Bernal (AB) stacking. The latter appears in three dimensional geometries describing the graphite natural configurations, which have received quite important deeper studies [24]. However, the (AA) stacking arrangement has been also synthesized and observed in few layer configurations including the bilayer hexagonal systems [25-28].

More recently, many efforts have been devoted to the study of the opening energy gaps in both single and multilayers of 2D honeycomb-structured materials using different approaches. In particular, methods based on exciton interactions [29], the application of the magnetic field [15,16], and curved geometries [15] have been well developed using various numerical calculation codes [30,31,32]. More precisely, the opening of the band gap in the graphene and the silicene has been investigated using first principle calculations by the help of quantum chemistry methods. In this context, the low-bucked structure in the silicene has been studied by turning on spin orbit interactions in the quantum spin Hall effect physics [33,34,35]. It is worth noting that experiments have also shown, recently, that the band gap can be opened up in a bilayer system by applying an electric field [36].

In this paper, we use Ab-initio calculations to discuss the band gap energy variation based on the effect of the inter-distance between two parallel layers of silicene separated by distance d. In particular, we consider a dynamic monolayer of the silicone located at distance d along the normal direction z forming with a static one a (AA) stacking arrangement. Here, we treat the planar and buckled geometries in (AA) arrangements, see figure 1. These buckled geometries involve two kinds of the configurations. The first one is associated with the usual (AA) configuration that keeps the same vertical distance between the two layers atoms, while the remaining one is obtained by reversing one the layer of the silicene. For both configurations, the starting distance is taking around the van der waals length. We show that the band gap can be opened by simply varying the distance, starting from a distance around the bond length of Van der Waals, between two parallel silicene for flat and buckled configurations. Furthermore, we study the stability between the buckled and the flat configuration in the mono and bilayer systems.

2. Building models and method

Before presenting the computation method, we first build the model. In fact, the systems we consider here are described by a static single layer of the silicone interacting with an extra dynamic layer. The latter is located at a distance d along the normal direction z forming with the first one a (AA) stacking arrangement as plotted in figure 1. The main object of this work is to see the effect of this extra layer of silicone by varying the distance d starting from a

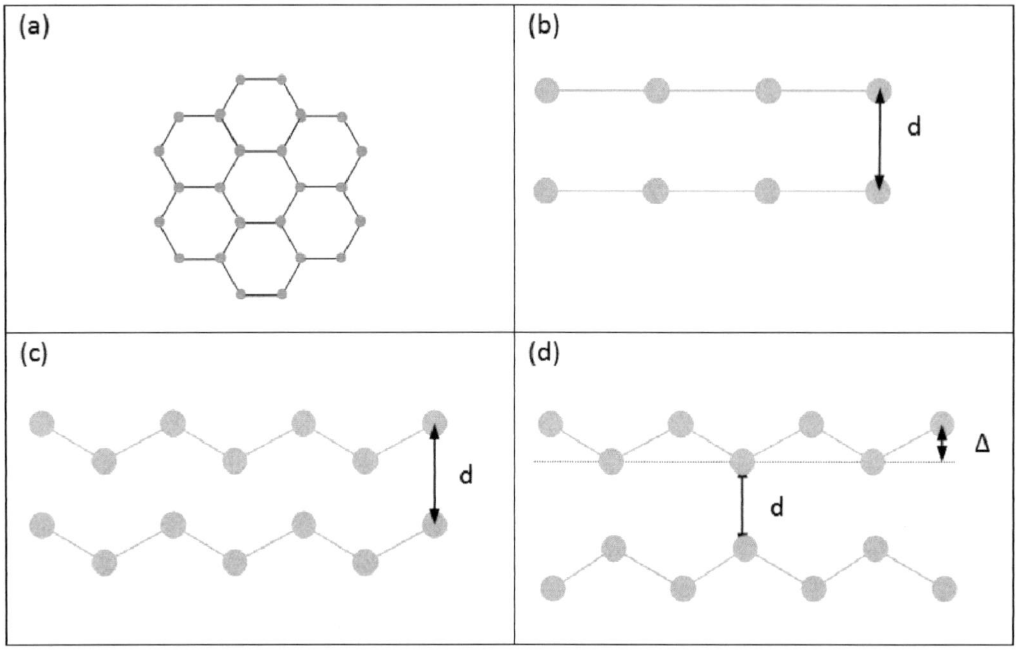

Figure 1. (a) hexagonal structure of silicene. (b) Two parallel flat silicene separated by distance d (c) buckled AA1 (d) buckled AA2 geometry.

distance around the van der Waals bond length. Here, we have performed ab intio calculations based on the all-electron full-potential local-orbital minimum-basis scheme FPLO9.00-34 [37,38]. This has been used to solve the Kohn-Sham equations using the scalar-relativistic scheme. The parameterization of the exchange-correlation energy has been done within the local density approximation [39]. To ensure a high accuracy in the performed computations, we have used both self-consistent criterions of the energy and the density together with a precision of 10^{-8}Ha and 10^{-6} respectively. To accurate the Brillouin zone integrations, we have used the tetrahedron scheme and the centered 48x48x1 K-point mesh. The valence basis state sets are (2s,2p,3s,3p,3d,4s,4p).

The lattice equilibrium parameters $a = b$ and the buckled parameter Δ has been determined from the relaxation method.

3. Numerical results and discussions

In this section, we discuss the distance variation effect on silicene like-bilayers with various geometries. First, we start with the flat case, and then we move to investigate the buckled geometries. A comparative between the studied geometries will be given and discussed.

3.1. Monolayer of the honeycomb-structured materials

Before performing the calculation corresponding to the effect of the inter-distance interaction on silicene like-bilayers, we have first checked the method for the case of the single layer of the silicene with flat and buckled geometries. In order to compute the density of states, we have first relaxed the planar lattice parameters ($a = b$) and the parameter Δ controlling the buckled geometries. Indeed, the total energy of the system is computed using the DFT method depending on the nucleon positions, hence one can minimize the energy functional with respect

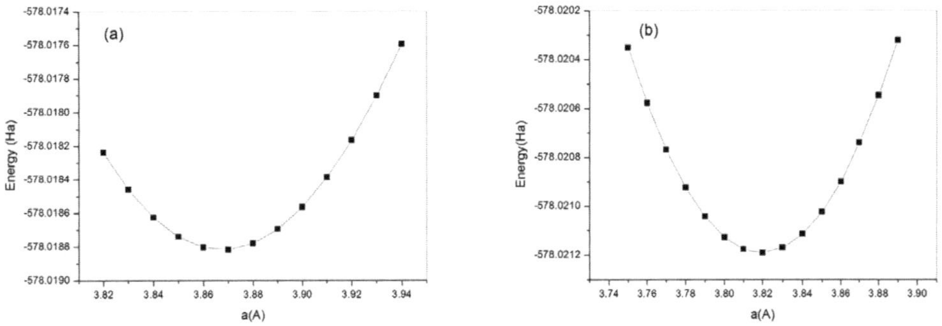

Figure 2. (a) Total energy as a function of lattice parameter a for flat (a) and buckled (b) silicene monolayer with LDA approximation.

Table 1. Lattice equilibrium positions for the silicene for flat and buckled for monolayers corresponding to LDA approximation.

	Flat	Buckled
$\Delta(\mathring{A})$	0	0.504
$a_{eq}(\mathring{A})$	3.87	3.82

to the positions. As usually, the total energy is a function of the lattice parameter a which has been calculated using method given in [40] based on FPLO9.00-34 with LDA approximations. Fitting the obtained result with respect to a generic polynomial of the second order, the lattice equilibriums and the buckled parameter of the silicene for both configurations are listed in the table 1. These values are comparable to those obtained recently in the literature [10]. Indeed, we can find that the buckled geometry is more stable than the flat one, see figure 2.

Considering the equilibrium positions, we have calculated the band structure around K-point for the silicene. The results are shown in the figure 3(a,b) for planar and buckled geometries respectively showing similar results obtained by other groups [10,19,41]. In particular, near the so-called Dirac points the system has a linear energy dispersion i.e

$$E = \pm h v_F k$$

where k and v_F are the momentum and Fermi velocity respectively. Linear dispersion behavior indicates that the charge carriers near the Dirac point become like massless Dirac obeying the relativistic Dirac-like equation. This is a good agreement allowing the validation of the modeling approximations based on FPLO9.00-34 code using the above mentioned parameterizations. It is recalled that the vanishing of the energy gap is intimately related to pz orbitals where the electrons can move freely. This is due to the fact that such orbitals are not interacting with extra ones. In order to reduce the high mobility of the electrons living on such orbitals, a new layer moving on the z direction should be introduced to build like bilayer systems. This extra layer may lead to new correlations between the electrons of pz orbitals placed in the two parallel layers of the silicene. At this level one may ask the following question. Do these correlations modify the band structure? This question will be addressed in the following subsection.

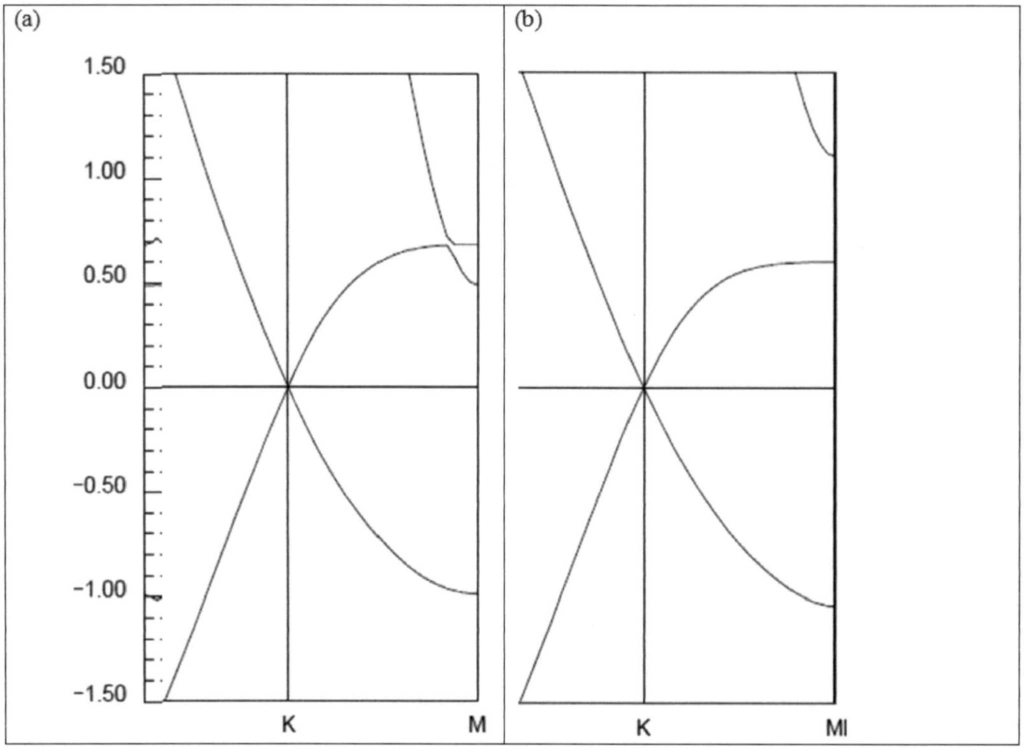

Figure 3. Monolayer Band Structure of silicon around K-point: (a) flat geometry and (b) buckled

3.2. Bilayers of silicene

In this subsection, we study the inter-effect of two layers system by changing the distance d between two parallel layers for planar and buckled geometries with the (AA) arrangement. For the buckled geometry, we distinguish two sorts of configurations. The first one corresponds to the usual (AA) configuration that keeps the same vertical distance between bilayer atoms, while the second one is obtained by reversing one buckled layer of the silicene. In the usual (AA) geometry, the buckled Si atoms of the top layer are faced with the buckled atoms living in the bottom layer (referred AA1). While in the reversed one, the buckled Si atoms of the top layer are faced to flat atoms of the bottom layers (referred AA2).

Varying the distance d, we have calculated the total energy in both configurations. These results are plotted in figure 4 (a,b and c) for buckled AA1, AA2 and flat geometries respectively. From these figures, we have found the equilibrium distance positions corresponding to the minimal energy for flat and two buckled geometries. The obtained results are listed in the table 2. From this table, we observe that we have two minima energies for flat geometry at the inter layers distances $4.2\mathring{A}$ and $2.4\mathring{A}$ which correspond to the Van der Waals and covalent bond length, respectively. However, it is shown clearly that, among these two minima the most stable minimum is related to the covalent distance $2.4\mathring{A}$ which is in a good agreement with the value reported in [42]. It follows that the minimal energy corresponding the minimal distance for two buckled geometries is less than the one associated with flat configuration. More precisely, the calculation indicates that the first buckled geometry is the most stable one.

Now we move to study the distance variation effect on the behavior of the opened energy gap and the electronic structure for the three possible configurations mentioned before. As we

Table 2. Equilibrium distance positions corresponding to the minimal energy for each configuration.

geometries	Equilibrium distance (Å)	Energy (Ha)
Buckled AA1	$d_{min} = 2.62$	$E_{min} = -1156.06276$
Buckled AA2	$d_{min} = 3$	$E_{min} = -1156.06249$
Flat	$d_{min1} = 2.4$ $d_{min2} = 4.2$	$E_{min1} = -1156.06155$ $E_{min2} = -1156.04003$

mentioned above, we can distinguish two zones that are related to the inter layer distance for the Van der Waals and covalent bond length. In fact, we focus, essentially, on the distance variation effect starting from the Van der Waals bond length. This choice is motivated from the fact that this bond is the weak one appearing in the two separated two layers system.

The present calculations show that the band energy gap can be opened up due to the influence of the variation of the distance d for the flat and buckled geometries for the inter layer distance staring from the Van der Waals bond length. This result, which has been obtained from the distance corresponding to a maximal gap, is shown in figure 5. In particular, we see that the band gap varies exponentially with the distance d. It has been also observed that the obtained values of the band energy of AA1 and flat geometries are bigger than the ones obtained in the case of the AA2 geometry. This difference of the energy gap values between the two configurations is due the interaction between two the silicene layers. This can be understood in terms of the distance between two opposite atoms, of the two layers, in both configurations. Indeed, for the AA1 configuration, the opposite atoms have the same distance allowing the maximum gap simultaneously. However, for theAA2 configuration, the opposite atoms do not have the same distance so the system cannot reach the maximum gap at the same distance.

To understand the origin of such an opening energy gap variation, we have calculated the band structure around K-point at the maximal gap distance for each geometry. The corresponding results are depicted in the figures 5(a,b,c,) for flat AA, buckled AA1 and AA2 respectively.

The distance variation shifts the pz orbital states to the valence band for small distance. For large distances, however, the system behaves like single layer of the silicene. It has been seen that the px and py orbitals are not changed. It is known that the electrons located at the valence band contribute to the local cohesion between neighboring atoms and they form localized states. Based on this observation, the variation of the distance has an influence on states of the pz orbital electrons. Taking small values of d, the pz states get stable situations forming a consistent bilayer system. This can be supported from the computed total energy depending on the distance d.

This stabilization can be understood in terms of an electronic structure transition in pz orbitals. It is a transition from delocalized to localized states allowing the opening of the energy gap. One possible explanation of this behavior can be given in terms of the coupling interaction between two parallel material planes caused by varying the interbilayer distance.

In the end of this work, we would like to make contact with many works dealing with the graphene like bilayers. Concerning the opening of energy gap, the buckled silicene like -bilayers share similar behaviors appearing in the case of the graphene only for the distance greater than Van der Waals bond length. The results obtained in this work may support the buckled silicene to be a potential candidate for the nanoscience applications.

Journal of Physics: Conference Series **491** (2014) 012006 doi:10.1088/1742-6596/491/1/012006

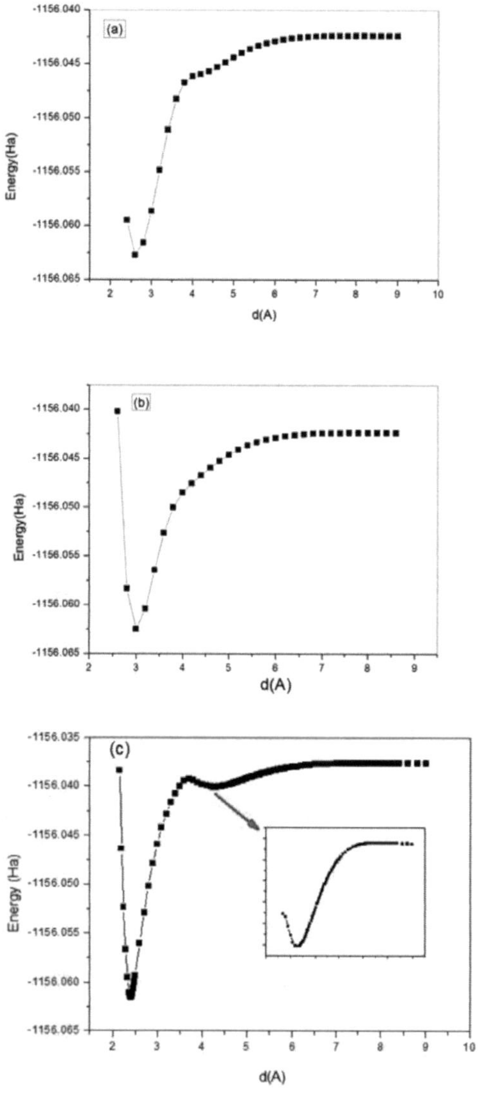

Figure 4. Total energy as function of distance d: (a) AA1 geometry, (b) AA2 geometry and (c) Flat geometry.

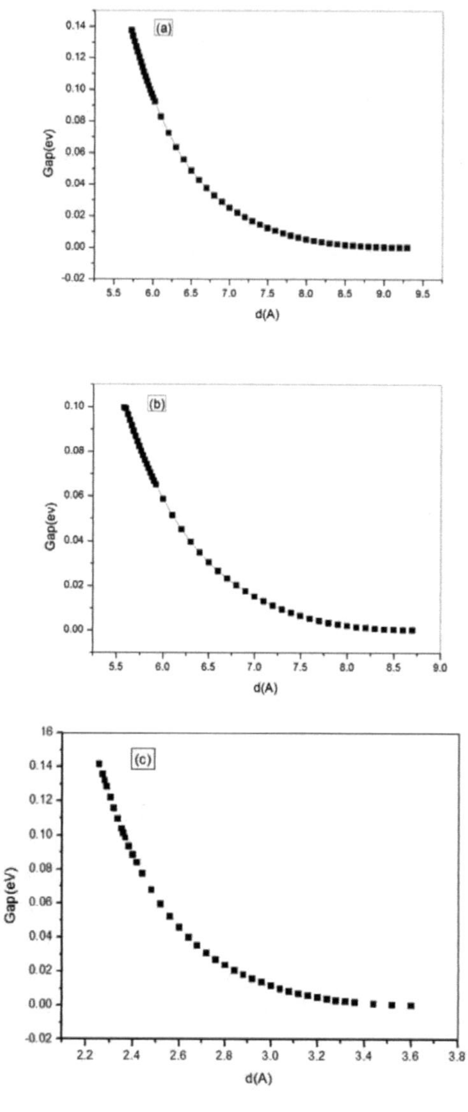

Figure 5. Band gap energy as function of distance: (a) for AA1 geometry , (b) AA2 geometry and (c) Flat geometry.

4. Conclusion

In this work, we have discussed the electronic band structure of two parallel planes of silicene having hexagonal geometries using DFT methods by help of first principal calculations based on FPLO9.00-34 code. We have presented an explicit computational study of a static single layer of silicene interacting with an extra dynamic one. The later is located at distance d along the normal direction z forming with the first one a (AA1) and reversed (AA2) stacking arrangements. It has been shown that band gap can be opened by simply varying the distance between two parallel layers of silicone only for the distance greater than Van der Waals bond length. The existence of the energy gap behavior is due to the transition from delocalized to localized states in pz orbitals. We believe that the obtained results present the first systematical study on silicene like bilayers.

References

[1] Novoselov K S et al 2004 *Science* **306** 666-667
[2] Geim A K and Novoselov K S 2007 *Nat. Mater.* **6** 183-191
[3] Castro Neto A H et al 2009 *Rev. Mod. Phys.* **81** 109-162
[4] Guzmán-Verri G G and Lew Yan Voon L C 2007 *Phys. Rev.*B **76** 07513191-100
[5] Kara A et al 2009 *J. Supercond. Novel Magn.* **22** 259-263
[6] Aufray B, Kara A et al 2010 *Appl. Phys. Lett.* **96** 183102
[7] Naji S, Belhaj A, Labrim H, Benyoussef A and El Kenz A 2012 *The Europ. Phys. Jour.*B **85** 373
[8] Naji S, Belhaj A, Labrim H, Benyoussef A and El Kenz A 2013 *Mod. Phys. Lett.* B accepted
[9] Drummond N D, Zolyomi V and Fal'ko V I arXiv:1112.4792.
[10] Behera H and Mukhopadhyay G 2010 *AIP Conf. Proc.* **1313** 152-155
[11] Cahangirov S, Topsakal M, Akturk E, Sahin H and Ciraci S 2009 *Phys. Rev. Lett.* **102** 236804
[12] Kara A, Enriquez H, Seitsonen A P, Lew Yan Voon L C, Vizzini S, Aufray B and Oughaddou H 2012 *Surf. Sci. Rep.* **67** 1-18
[13] Takeda K and Shiraishi K 1994 *Phys. Rev. B* **50** 14916
[14] Zhang Y et al 2005 *Nature* **438** 201
[15] Castro Neto A H, Guinea F, Peres N M R, Novoselov K S and Geim A K 2009 *Rev. Mod. Phys.* **81** 109
[16] Vozmediano M A H and Guinea F arXiv:1108.0580.
[17] Feng B, Ding Z, Meng S, Yao Y, He X, Cheng P, Chen L and Wu K arXiv:1203.2745.
[18] Houssa M, Pourtois G, Afanasev V V and Stesmans A 2011 *Appl. Phys. Lett.* **97** 112106
[19] Lebegue S and Eriksson O 2009 *Phys. Rev.* B **79** 115409
[20] Dzade N Y, Obodo K O, Adjokatse S K, Ashu A C, Amankwah E, Atiso C D, Bello A A, Igumbor E, Nzabarinda S B, Obodo J T, Ogbuu A O, Femi O E, Udeigwe J O and Waghmare U V 2010 *J. Phys.Condens. Matt.* **22** 375502
[21] Jose D and Datta A 2011 *Phys. Chem. Chem. Phys.* **13** 7304
[22] Berashevich J and Chakraborty T arXiv:1107.5299
[23] Mak K F, Shan J, Tony F and Heinz 2010 *Phys. Rev. Lett.* **104** 176404
[24] Charlier J C, Gonze X and Michenaud J P 1994 *Europhys. Lett.* **28** 403
[25] Norimatsu W and Kusunoki M 2010 *Phys. Rev.*B **81**161410(R)
[26] Lee J K, Lee S C, Ahn J P, Kim S C, Wilson J I B and John P 2008 *J. Chem. Phys.* **129** 234709
[27] Liu Z, Suenaga K, Harris P J F and Iijima S 2009 *Phys. Rev. Lett.* **102** 015501
[28] Ohta T, Bostwick A, Seyller T, Horn K and Rotenberg R 2006 *Science* **313** 951
[29] Roldan R, Fuchs J N and Goerbig M O 2010 *Phys. Rev.* B **82**, 205418
[30] Kresse G and Furthmuller J 1996 *Phys. Rev.* B **54** 11169-11186
[31] Akhukov M A, Fasolino A, Gornostyrev Y N and Katsnelson M I 2012 *Phys. Rev.* B **85** 115407
[32] Konschuh S, Gmitra M, Kochan D and Fabian J arXiv:1111.7223.
[33] Ezawa M arXiv:1203.0705.
[34] Ezawa M arXiv:1202.1357.
[35] Cheng Liu C, Feng W and Yao Y 2011 *Phys. Rev. Lett.* **107** 076802
[36] Kamal C arXiv:1202.2636.
[37] http://www.fplo.de.
[38] Koepernik K and Eschrig H 1999 *Phys. Rev.* B **59** 1743
[39] Perdew J P and Wang Y 1992 *Phys. Rev.* B **45** 13244
[40] Fenman R P 1939 *Phys. Rev.* **56** 4
[41] Sahin H et al 2009 *Phys. Rev.* B **80** 155453
[42] Tritsaris G A, Kaxiras E, Meng S and Wang E 2013 *Nano Lett.* **13** 2258

Silicene on substrates: interaction mechanism and growth behavior

H Liu[1], J Gao[1,2], J Zhao[1,2*]

[1]Key Laboratory of Materials Modification by Laser, Ion and Electron Beams (Dalian University of Technology), Ministry of Education, Dalian 116024, China

[2] Beijing Computational Science Research Center, Beijing 100084, China

E-mail: zhaojj@dlut.edu.cn

Abstract. Silicene, a monolayer of silicon atom assembling in a honeycomb lattice, has attracted more and more attention due to its outstanding electronic properties. The recently successful synthesis of silicene on several metal surfaces takes a big step towards the utilization of silicene in the future microelectronic devices. On the roadmap for the applications of silicene, two critical issues have to be addressed: (1) how to improve the quality of silicene; (2) how to preserve the extraordinary electronic properties of silicene. These two problems can be solved by deeply understanding the substrate effect on silicene. In this review, we systematically discuss the substrate effect on the atomic structure and electronic properties of a silicene sheet as well as the growth behavior of silicene on Ag surface, which are important for both fabrication and application of silicene.

1. Introduction

As early as in 1994, the theoretical possibility of stage corrugation in Si analogs of graphite had already been explored.[1] But until 2009, two-dimensional (2D) silicon monolayer possessing a hexagonal honeycomb lattice with a buckled height of about 0.44 Å was confirmed to be stable by phonon dispersion and *ab initio* finite temperature molecular dynamics calculations.[2,3] Such single atom thickness of 2D sheet silicon was thereafter named by silicene[4-34], which attracts more and more attentions recently. Most of all, silicene possesses outstanding electronic properties similar to that of graphene.[2-6,35-37] The band structure of silicene exhibit linear dispersion around Fermi level, forming the famous "Dirac cone" at the symmetric point K in the reciprocal space. This band characteristic indicates that the charge carriers in silicene sheet behave like massless Dirac fermions in a small energy range around the Fermi level E_F. Interestingly, the velocity of charge carriers near Fermi level in silicene was estimated to be $10^5 \sim 10^6$ m/s,[2] comparable to that in graphene.[37-39] Both first-principle and tight-binding calculations show that silicene has a much larger band gap (about 1.55 meV) at the Dirac points opened by the effective spin-orbit coupling than graphene.[7,8] Therefore the quantum spin Hall effect can be observed in silicene in an experimentally accessible temperature regime, which may lead to some novel silicene-based spintronic devices. Moreover, first-principles calculations predicted that a tunable gap can be opened up in silicene at the Fermi level via hydrogenation,[9-11] fluorination,[12] or applying electric field.[13,14] Most attractively, easy integration into the current Si based technology makes silicene a potential candidate for microelectronic devices.

To realize the utilization of silicene, the challenge for fabrication of silicene must be overcome. Since there is no silicon allotrope with layered structure analogous to the graphite in nature, the mechanical stripping method for fabricating graphene[35] and many other 2D materials[40] does not work for silicene. Meanwhile, the silicon oxide (SiO_2) is a kind of hard insulator. Thus, the method of reduction of graphene oxide[41] can't be applied to silicene either. Up to now, the major method for synthesizing silicene is epitaxial growth on metal substrates. Before the synthesis of large-scale silicene sheet, silicene nanoribbion was successfully fabricated on Ag(001)[42] and Ag(110)[43-45] surfaces. Afterwards, Lalmi et al. reported epitaxial growth of silicene on Ag(111) substrate in 2010.[15] The lateral Si-Si distance in the $(2\sqrt{3}\times 2\sqrt{3})R30^\circ$ superstructure observed by them was 0.19 ± 0.01 nm, shorter than the Si-Si bond lengths ($0.22 \sim 0.24$ nm) from previous theoretical prediction[3] and experimental observation in silicon nanoribbons[39,40]. The difference between the lateral Si-Si distance and the Si-Si bond lengths was attributed to the large buckling of silicene on Ag(111) surface.[32] More recently, several experiments were conducted to repeat the epitaxial growth of silicene on Ag(111) surface and a variety of superstructures were observed, including (4×4), $(\sqrt{13}\times\sqrt{13})R13.9^\circ$, $(\sqrt{7}\times\sqrt{7})R19.1^\circ$, $(2\sqrt{3}\times 2\sqrt{3})R30^\circ$ with respect to Ag(111)[17-20,33,46]. The average Si-Si distances in these superstructures are 0.22 ± 0.01 nm, in agreement with that in nanoribbon. Our first-principles simulations explained the superiority of Ag as substrate for the growth of silicene.[21] Firstly, Ag(111) surface can stabilize sp^2 hybridized silicene via moderate interaction between silicene and surface metal atoms. Secondly, Ag(111) surface interacts with silicene almost homogeneously and thus the local tension in silicene during the growth process can be rather small.

In addition to Ag, efforts also have been made to synthesize silicene on other substrates. For example, silicene was also successfully synthesized on zirconium diboride (ZrB_2) thin films grown on Si wafers through epitaxial growth[22] and Ir(111) surface by directly depositing silicon atoms and annealing the sample at 670 K[23]. The successful synthesis of silicene ribbons and sheets supplies the precondition for utilizing silicene in nanoscale materials and devices. The atomic structures, electronic properties and synthesis of silicene have been summarized in previous reviews.[47-49]

Though free-standing silicene exhibits excellent electronic properties, during the growth process of silicene or in the practical utilization of silicene in microelectronic devices, it always occurs along with substrates. Thus, the substrate effects are significant and have attracted some recent efforts.[14,16,24-30,34] In this review, we will briefly summarize our current understanding on the substrate effects for silicene and the growth behavior of silicene on Ag(111) surface.

2. Interaction mechanism between silicene and substrates

2.1 silicene on insulating substrates

In the future microelectronic devices made of silicene, silicene has to be supported on some insulating substrates, similar to the cases of graphene[50-58] and MoS2 sheets[59-61]. Thus, the effect of inert substrates on the electronic and transport properties of silicene is an important issue for the practical applications of silicene. Houssa et al. proposed that a silicene flake could preserve the semimetallic behavior when it is embedded between ultrathin AlN layers.[24] The ultrathin AlN was shown to be insulating with an energy gap of about 4.6 eV. Silicene flake weakly interacts with AlN layers via vdW forces. Recent *ab initio* calculations suggested that the excellent electronic properties of silicene can be preserved on hexagonal boron nitride (*h*-BN) monolayer and hydrogenated SiC(0001) surface[14,25,26] It was found that silicene retain its geometry structure with a distance of about 3 Å from the substrates (Figure 1). The cohesive energies between silicene and these substrates are about 0.07~0.09 eV per silicon atom, which can also be viewed as van der Waals (vdW) type.

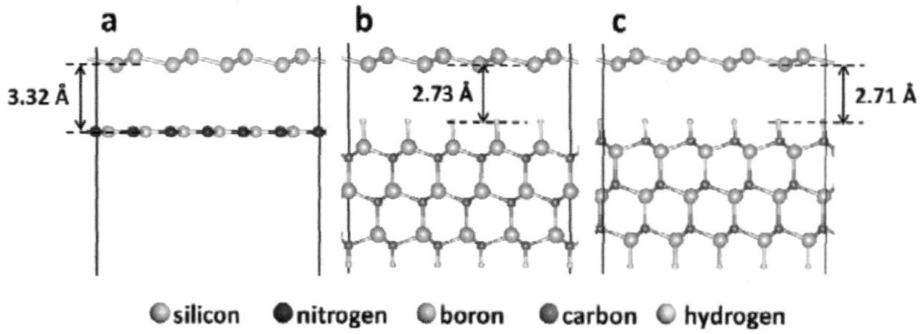

Figure 1. Atomic structures of silicene on: (a) *h*-BN; (b) Si-SiC; (c) C-SiC.[25]

As shown in Figure 2b, c, the Dirac cone of silicene was nearly preserved on the *h*-BN monolayer and hydrogenated Si-terminated SiC(0001) surface (Si-SiC). Therefore, *h*-BN and Si-SiC can provide effective mechanical support for silicene without disturbing its geometry and electronic properties. On the contrary, silicene on the hydrogenated C-terminated SiC(0001) surface (C-SiC) becomes metallic (Figure 2d). For silicene/BN (Figure 2f) and silicene/Si-SiC (Figure 2g), the partial charge densities in the energy range of E_F~E_F−0.2 eV concentrate on silicene, similar to that in free-standing silicene (Figure 2e). While for siliene/C-SiC, the partial charge densities distribute widely in silicene and the substrate (Figure 2h), implying that the Fermi level has already entered the vicinity of valence band maximum (VBM) of C−SiC.

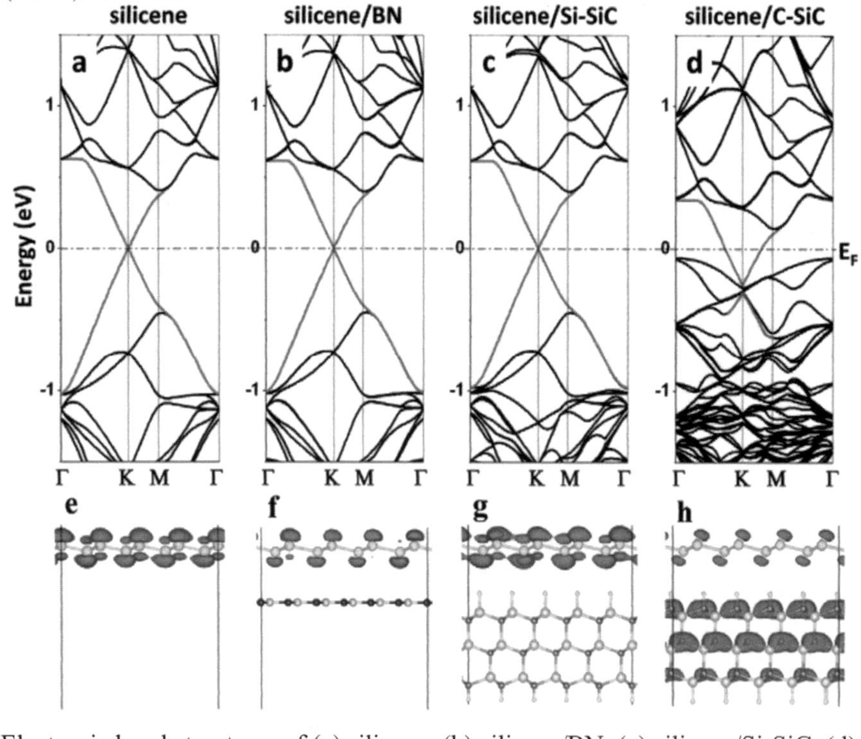

Figure 2. Electronic band structures of (a) silicene, (b) silicene/BN, (c) silicene/Si-SiC, (d) silicene/C-SiC and partial charge densities of (e) silicene (f) silicene/BN, (g) silicene/Si-SiC, (h) silicene/C-SiC. The red lines highlight the π-bands of silicene in different configurations. The partial charge densities are in the energy range of E_F~E_F−0.2 eV.[25]

Different substrate effects on the electronic property of silicene can be explained by the work functions of substrates.[25] By setting the vacuum energy as zero, the VBM is the negative value of work function. As shown in Figure 3a, the VBM of silicene is slightly lower than the VBM of C-SiC. The approaching of the two energy levels results in certain interaction between silicene and substrate. As a consequence, some new electronic states arise between the two VBMs due to the strong electron coupling after silicene contacting with C-SiC. Thus, the Fermi level of the silicene/C-SiC hybrid system lies between the VBM of original silicene and the VBM of original C-SiC (Figure 3b). Eventually, the Dirac cone of silicene is submerged under the Fermi level (Figure 3b) and slightly disturbed due to the electron coupling. On the contrary, for the other two substrates, the VBM of silicene locates in the gap of the substrates that is far from the VBMs of the substrates. The electron coupling between silicene and the substrates is too weak to create any new electronic states. Consequently, the Fermi levels for silicene/BN and silicene/Si-SiC still cross the Dirac points of silicene.

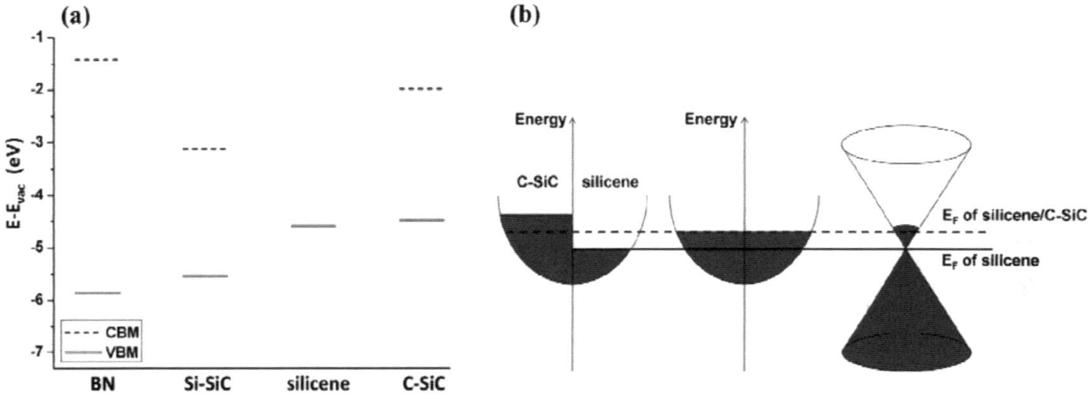

Figure 3. (a) Conduction band minima (CBM) and valence band maximum (VBM) of BN monolayer, Si−SiC substrate, silicene, and C−SiC substrate with respect to the vacuum level (set as zero energy). The values of CBM and VBM of silicene are exactly identical. (b) Schematic for the origin of metallic behavior in silicene/C−SiC hybrid system. Blue areas present the occupied electronic states.[25]

2.2 Silicene on metal substrates

Different from the insulating substrates, silicene sheets forms various superstructures on metal surfaces due to stronger silicene-substrate interactions.[17-20,22,23,46] The buckled height is about 1 Å for silicene on Ag(111) surface. As discussed above, some insulating substrates with appropriate work function can preserve the Dirac cone of silicene and thus the outstanding electronic properties. In the case of metal substrates, there is a fierce debate about whether the Dirac cone can be preserved on Ag(111) surface.[16,26-30] In 2012, using angular-resolved photoelectron spectroscopy (ARPES) measurements, linear dispersion was observed in the (3 × 3) silicene on (4 × 4) Ag(111) superstructure (refer as (4 × 4) superstructure hereafter) with the Dirac cone about 0.3 eV below the Fermi level.[16] It was proposed that the linear dispersion came from silicene and a Fermi velocity of 1.3×10^6 ms^{-1} was estimated. However, combining spectroscopic signatures of quantum Hall effect by the Landau quantization under a magnetic field and density functional theory calculations, Takagi and co-workers proposed that silicene would no longer be a two-dimensional Dirac fermion system when synthesized on the Ag(111) surface due to substrate-induced symmetry breaking.[27] Figure 4a shows the scanning tunneling microscopy (STM) image of a (4 × 4) superstructure. The amplificatory STM image is shown in Figure 4b. Scanning tunneling spectroscopy (STS) spectra was used to measure the Landau levels. For the magnetic field perpendicular to the sample between zero and 7 T, characteristic structures due to the Landau quantization were not found (Figure 4c), which implies the silicene loses its characteristic of Dirac fermion system. In contrast, Landau levels appear in the highly-oriented

Journal of Physics: Conference Series **491** (2014) 012007 doi:10.1088/1742-6596/491/1/012007

pyrolytic graphite (HOPG) (Figure 4d), which can be reasonably rationalized by a combination of massless and massive Dirac fermions.[62-66]

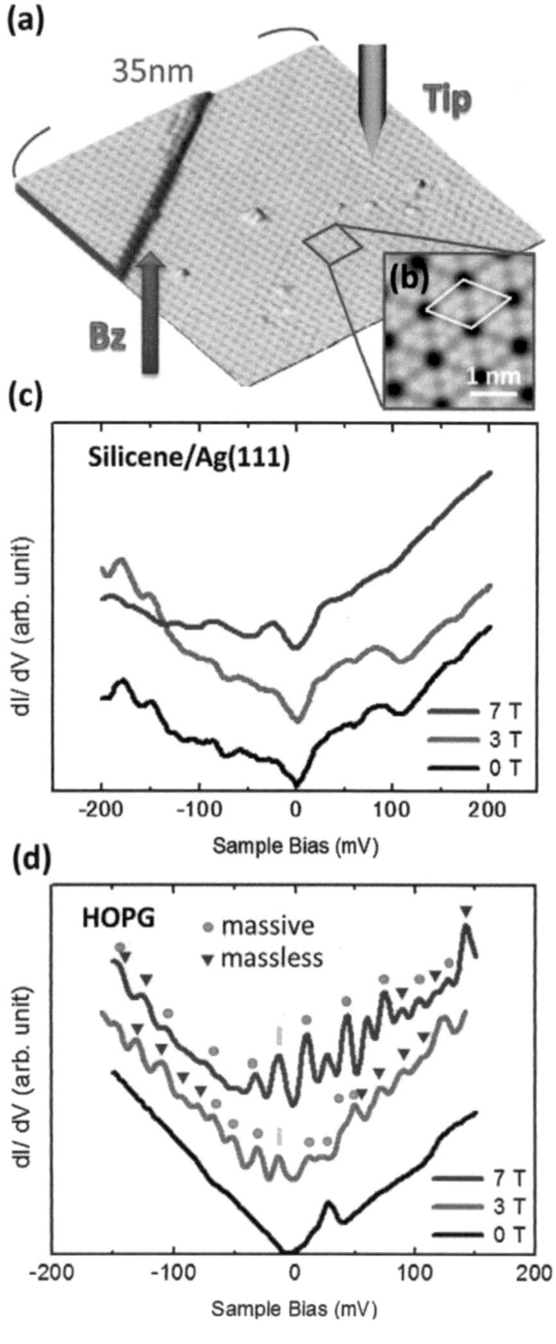

Figure 4. (a) STM image of large area of silicene (sample bias voltage V_S = -0.70 V and tunneling current I_t = 0.19 nA. The image size is 35 × 35 nm^2). (b) High resolution STM image of the 4 × 4 silicene (V_S = +0.50 V and I_t = 0.30 nA, 3.65 × 3.65 nm^2). The unit cell is shown by the white rhombus. (c) STS spectra of silicene for various magnetic fields perpendicular to the sample surface, B_Z. (d) STS spectra of HOPG for various B_Z. The purple triangles and green circles show the peaks originating from the LLs of massless and massive Dirac fermions, respectively. The LL with n = 0 is marked by the yellow bar and the LL with n = 1 of massive Dirac fermions is not clearly resolved in 3 T due to low magnetic field.[27]

Detailed analysis of Kohn-Sham orbitals from first-principles calculations of the (4 × 4) superstructure along the Γ–K direction showed the linear band observed by ARPES experiment may correspond to the sp band of Ag substrate.[26] Almost at the same time, another group independently demonstrated that the Dirac cone in silicene on Ag(111) surface was destroyed due to the distortion of silicene on Ag surface and strong band hybridization at silicene/Ag(111) interface, by unfolding bands from supercell calculations.[28] They proposed that the linear dispersion observed in angular-resolved photoemission spectroscopy[16] indeed comes from the Ag substrate, instead of silicene. Very recent first-principles calculations show that the linear dispersion observed in experiment originates from the hybridization states between silicene and Ag(111) surface, neither from silicene nor from Ag sp bands.[29] The reconstructed (3 × 3) silicene without Ag substrate opens a 0.3 eV band-gap at the pristine Dirac cone (red dots in the left panel of Figure 5) due to structural modification induced by the Ag substrate. Comparing the unfolded band of reconstructed (3 × 3) silicene (red dots in the left panel of Figure 5) with the unfolded band of silicene/Ag (red dots in the right panel of Figure 5), no one to one correspondence can be found between them. Thus, a strong hybridization occurs between silicene and Ag substrate. The surface states of Ag have no contribution to the linear dispersion in absence of silicene (Figure 6b). When silicene is present, however, the linear band appears again (Figure 6c). This coincides with the experiment observation.[16] Moreover, the linear dispersion from bulk Ag states is absent at the band structure of surface Ag states. Therefore, Cahangirov et al. proposed that it is more reasonable to attribute the observed linear bands to the hybridized states of silicene and Ag rather than bulk Ag state.[29] On the contrary, Avila and the co-workers recently revisited the electronic structures of the 3×3 silicene phase on Ag(111) surfaces by using a compiled set of LEED, CLs and ARPES data, pronouncing that the linear dispersion still comes from silicene rather than Ag substrate.[30]

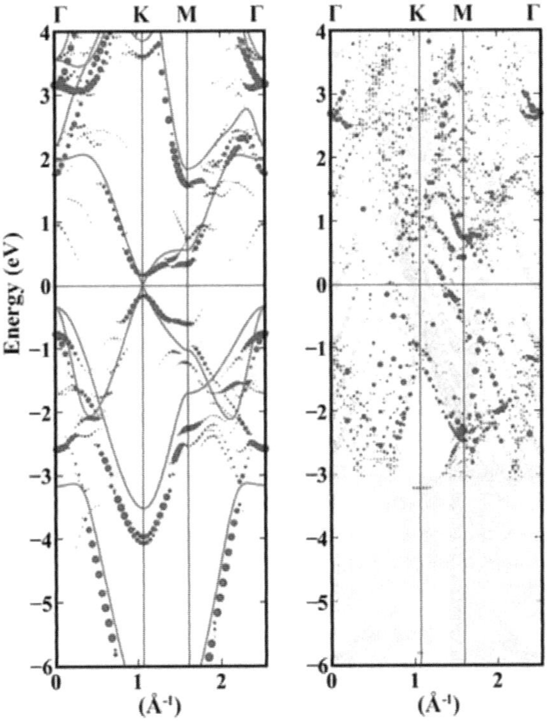

Figure 5. Left panel: bands of reconstructed (3 × 3) silicene in the absence of Ag substrate (unsupported silicene) unfolded to BZ of (1 × 1) silicene are shown by red dots. The radii of dots correspond to the weight of unfolding. The band structure of ideally buckled silicene is shown by green lines. Right panel: unfolded band structure of silicene on Ag. Red dots correspond to states with significant contribution from silicene.[29]

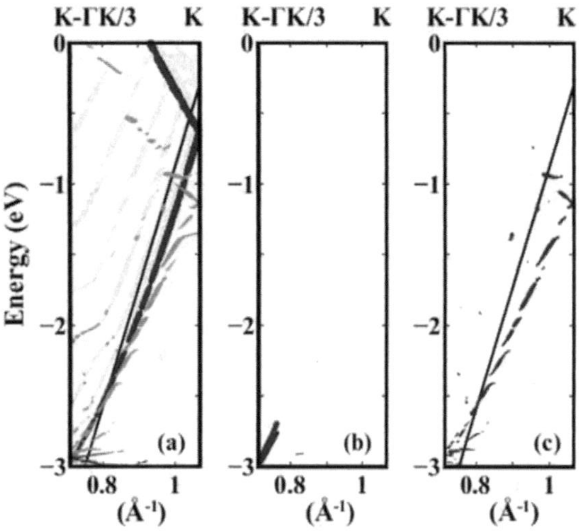

Figure 6. (a) Band structure of silicene with 11 Ag layers around K point. Blue and orange dots correspond to bulk Ag states and hybridized silicene/Ag states. Gray dots come from foldings of finite slab and can be ignored. Contribution of three Ag layers at the surface in the (b) absence and (c) presence of silicene.[29] The solid black lines in (a) and (c) are representative sketches of the experimental data.[16]

Besides the characteristics of electronic band structures, understanding the interaction mechanism between silicene and metal surface is crucial for growth of high quality silicene. Among these metal substrates, Ag(111) surface is widely used. Hence, it may serve as a representative to expatiate the interaction mechanism between silicene and metal substrates. A silicon clusters Si_{24} composed of seven hexagonal rings, which can be viewed as a small patch of silicene, was constructed to explore the interaction between silicon atoms and Ag(111) surface.[21] Upon relaxation, Si_{24} cluster retains its geometry on Ag(111) surface (Figure 7a). The local density of states (LDOS) for Si_{24}@Ag(111) (Figure 7b) shows that the 3p states of Si atoms couple with the 4d states of Ag atom in the vicinity of the Fermi level, implying the hybridization between Si p_z electrons and Ag 4d electrons. Such hybridization may also explain why silicene loses its Dirac fermion characteristics on Ag(111) surface. The charge differential densities of Si_{24}@Ag(111) shown in Figure 7d illustrate that all Si atoms at both periphery and inner regions of Si_{24} cluster interact evenly with Ag(111) surface. Moreover, the on-site charges of Si atoms in Si_{24}@Ag(111) system (Figure 7c) are nearly homogeneous (range between -0.13 $|e|$ and -0.08 $|e|$). This finding coincides with the small difference of adsorption energies for Si atom on different sites of Ag(111) surface (less than 0.031 eV)[21]. Thus, the local tensions in Si_{24} cluster and other silicene patches induced by the Ag(111) surface should be small, which is surely beneficial for continuous growth of high-quality silicene. In contrast, the difference of adsorption energies for Si atom on different sites of Rh(111) surface is calculated to be 0.306 eV[21], much larger than that for Ag(111) surface. In short, moderate and homogeneous interaction between Si atoms and Ag atoms makes Ag(111) surface a suitable substrate for epitaxial growth of silicene.

Journal of Physics: Conference Series **491** (2014) 012007 doi:10.1088/1742-6596/491/1/012007

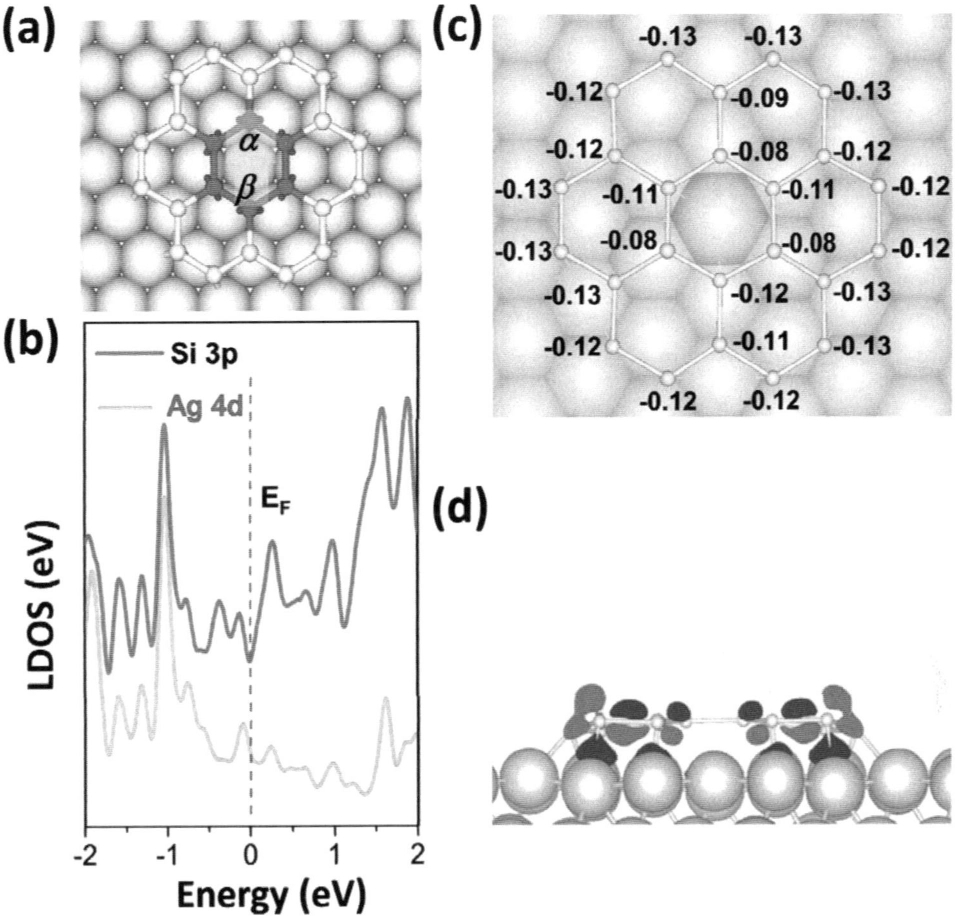

Figure 7. (a) Atomic structures, (b) local density of states (LDOS), (c) on-site Mulliken charges (|e|), (d) charge differential densities of Si_{24}@Ag(111) system. The LDOS comes from the central six silicon atoms (3p orbital) highlighted by red balls as well as the Ag atoms (4d orbital) highlighted by green ball right underneath the Si hexagon. In (d), red (blue) zone loses (gains) charge.[21]

Due to the moderate silicene-substrate interaction, silicene sheets can be rather stable on Ag(111) surface and form some superstructures, including (3×3) silicene on (4×4) Ag(111) surface, ($\sqrt{7}\times\sqrt{7}$) silicene on $2\sqrt{3}\times2\sqrt{3}$ Ag(111) surface, ($\sqrt{7}\times\sqrt{7}$) silicene on $\sqrt{13}\times\sqrt{13}$ Ag(111) surface.[21] The atomic structures and simulated STM images of these silicene@Ag(111) superstructures are shown in Figure 8. Note that the simulated STM images agree well with experimental ones[16-20]. After geometry optimization, silicene sheets retain the honeycomb structure with a buckled height of about 1 Å. Only the topmost silicon atoms in the silicene sheets (highlighted by red balls) can be seen in the STM images. The stability of different silicene superstructures is characterized by the binding energy, defined as

$$E_b = E_{si} + E_{sub} - E_t \qquad (1),$$

where E_{si} is the energy of silicene, E_{sub} is the energy of Ag substrate, E_t is the energy of silicene/Ag system. According to the computed binding energy, we can compare the stability of three types of silicene superstructures on Ag(111) surface. The type I, i.e., (3×3) silicene on (4×4) Ag(111) with a binding energy of 0.368 eV per silicon atom, is the most stable one. The binding energies of other two silicene phases (($\sqrt{7}\times\sqrt{7}$) silicene on ($2\sqrt{3}\times2\sqrt{3}$) or ($\sqrt{13}\times\sqrt{13}$) Ag(111) surface) are very

Journal of Physics: Conference Series **491** (2014) 012007 doi:10.1088/1742-6596/491/1/012007

closed, which are 0.336 eV and 0.332 eV per Si atom, respectively. This may help understand why the type I is the most frequently observed phase of silicene on Ag(111) surface in experiments.[30]

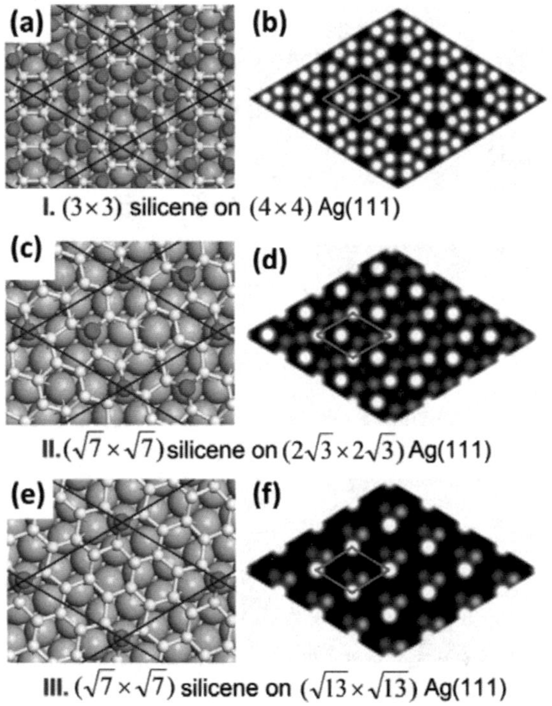

Figure 8. Atomic structures and simulated STM images of three types of silicene@Ag(111) superstructures. (a), (c), (e) are geometry of silicene@Ag(111) superstructures with topmost Si atoms highlighted by red balls; (b, d, f) are the STM images with a constant height of 2 Å to the topmost Si and a bias voltage of −1.5 V.[21]

Ab initio molecular dynamics (AIMD) simulations were also carried out to assess the thermal stability of (3 × 3) silicene superstructure on (4 × 4) Ag(111) surface.[21] The snapshots from AIMD simulation of this superstructure are shown in Figure 9. At 500 K (which is about the growth temperature of silicene), no topological defect was ever generated during the entire simulation time up to 7.5 ps, indicating high thermal stability of silicene on Ag(111) surface. At 900 K, some defects arose in silicene after 0.8 ps (Figure 9c), but were spontaneously recovered very quickly.

For comparison, a silicene@Rh(111) superstructure ((4 × 4) silicene on (6 × 6) Rh(111) surface) was constructed. After DFT relaxation, silicene keeps its honeycomb geometry on Rh surface (Figure 9d). However, after AIMD simulation of only 0.7 ps at 500 K, two Si-Si bonds were broken (Figure 9e). Afterwards, the silicene sheet became amorphous after 2.3 ps at 500 K (Figure 9f). As discussed above, the diffusion barrier of silicon monomer on Rh(111) surface is about ten times of that on Ag(111) surface. Therefore, the thermal instability of silicene monolayer on Rh(111) surface can be ascribed to the larger local energy difference of Si atom on Rh surface. Again, comparative AIMD simulations demonstrate that the superiority of Ag as a substrate for silicene growth originates from the intermediate and homogeneous interaction between Si atoms and Ag atoms.

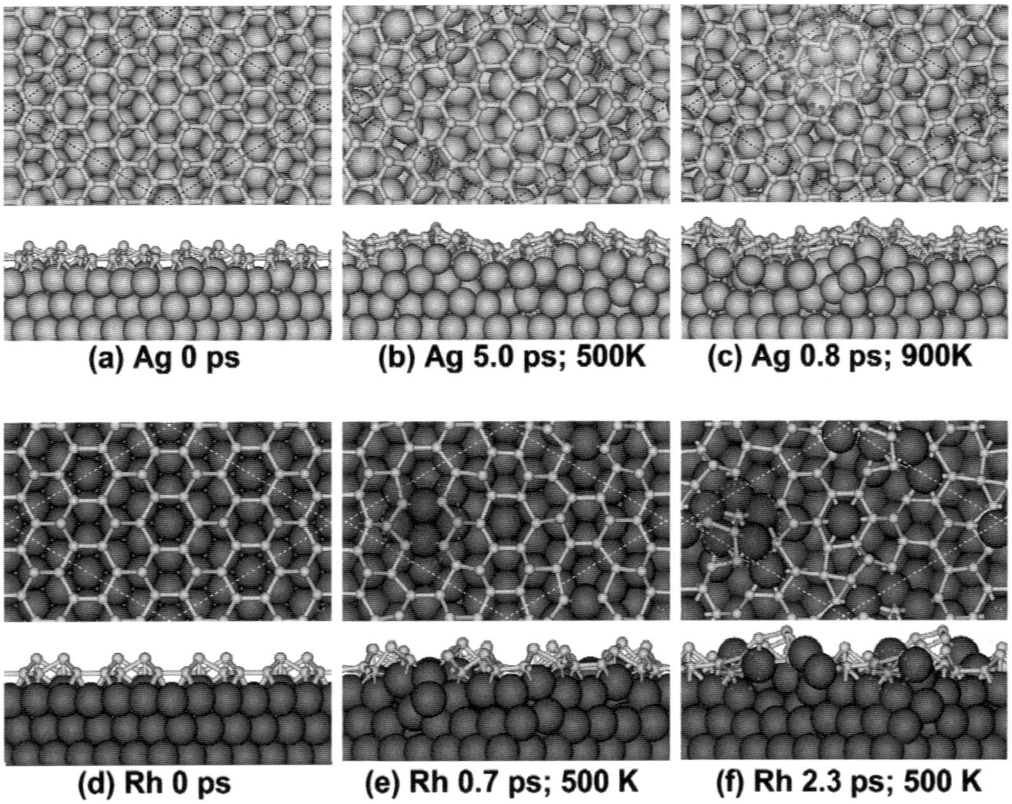

Figure 9. Snapshots from AIMD simulation of (5 × 5) silicene monolayer on (7 × 7) Ag(111) surface with 3.8% lattice mismatch: (a) initial configuration, (b) structure after 5.0 ps at 500 K, (c) structrue after 0.8 ps at 900 K. Snapshots from AIMD simulation for (4 × 4) silicene on (6 × 6) Rh(111) surface with 3.7% lattice mismatch: (d) initial configuration, (e) structure after 0.7 ps at 500 K, (f) structure after 2.3 ps at 500 K.[21]

3. Growth behavior of silicene

Typically, the process for epitaxial growth of silicene on metal surface involves three stages: (i) silicon clusters are deposited on metal surface; (ii) silicon monomers or dimers aggregate into many small silicon clusters, but most of them dissociate into monomers, dimers and small pieces again; (iii) continuous growth of silicon clusters beyond nucleation size to cover the whole metal surface. Among the three stages, stage (ii), the nucleation of silicene, is crucial for understanding and controlling silicene growth on metal surface. Thus, it is important to explore the geometries and energies of small silicon clusters on metal surface in order to understand the growth mechanism of silicene, similar to the case of graphene.[67-69]

In vacuum, small silicon clusters prefer supercluster configuration composed of Si_6, Si_9, and Si_{10} subunits (Figure 10a)[70-72]. For comparison, we have also considered small planar Si_N clusters (N=6, 10, 13, 16, 19, 22, 24) as aggregates of six-membered rings (6MRs). Upon relaxation, these Si_N clusters can't keep planar in vacuum and transform into severely buckled configurations (see Figure 10b).[21] This is because silicon atom doesn't like sp^2 hybridization in vacuum. Therefore, homogeneous nucleation in vacuum is impossible for synthesis of silicene and a completely wet substrate (such as Ag) is needed.

During the epitaxial growth of silicene on Ag(111) surface, a silicene sheet forms by the continuous growth of small silicon cluster. Thus, elucidating the behavior of small silicon clusters on Ag(111) surface is significant for the growth of silicene. Figure 10c shows silicon monomer, dimer

Journal of Physics: Conference Series **491** (2014) 012007 doi:10.1088/1742-6596/491/1/012007

and 6MR-based silicene-like 2D clusters on Ag(111) surface. Except for Si_6 and Si_{13}, the planar honeycomb structures can be preserved on Ag(111) surface, which may act as the nucleation center during the growth of silicene.

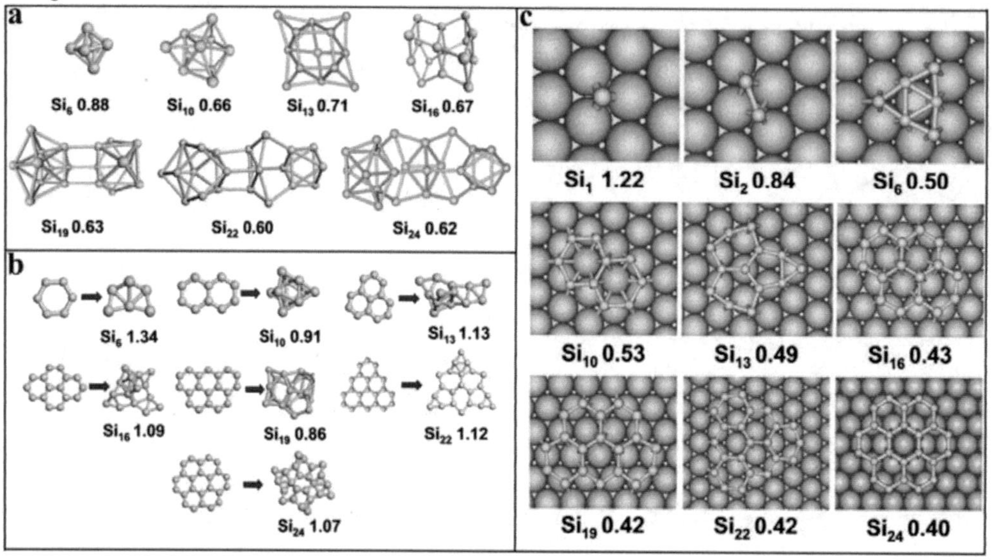

Figure 10. (a) Ground state structures and their formation energies (eV per atom) for Si_N clusters (N=6, 10, 13, 16, 19, 22, 24) in vacuum; (b) planar Si_N clusters composed of hexagonal rings (left) and the corresponding structures after relaxation in vacuum (right); (c) geometries and formation energies (eV per Si atom) of 2D Si_N clusters on Ag(111) surface.[21]

Due to metal passivation effect (Figure 11), the formation energies of clusters decrease after they are adsorbed on Ag(111) surface. More interestingly, on Ag(111) surface, the planar Si_N clusters have lower formation energies than the corresponding ground state Si_N cluster in vacuum (Figure 11). Thus, on Ag surface, silicon clusters prefer 2D planar structures to 3D structures, which implies that Ag(111) surface is a wet substrate. Moreover, the formation energy for silicene-like 2D cluster decreases as the number of hexagonal ring increasing (Figure 11), which means that the aggregation of silicon atom is energetically favorable.

Another important factor for the growth of silicene is the diffusion barrier for silicon atom on Ag(111) surface. NEB calculation showed that the diffusion barrier for silicon monomer on Ag(111) is as low as 0.031 eV, implying easy diffusion of silicon atom on Ag surface.[21] As a consequence, after depositing silicon atoms on Ag(111) surface, these silicon atoms would quickly aggregate into small planar clusters, overcoming a small barrier. Small patches of silicene will occur at some critical cluster size and grow into large silicene sheet by continuously adsorbing silicon atoms on the Ag surface.

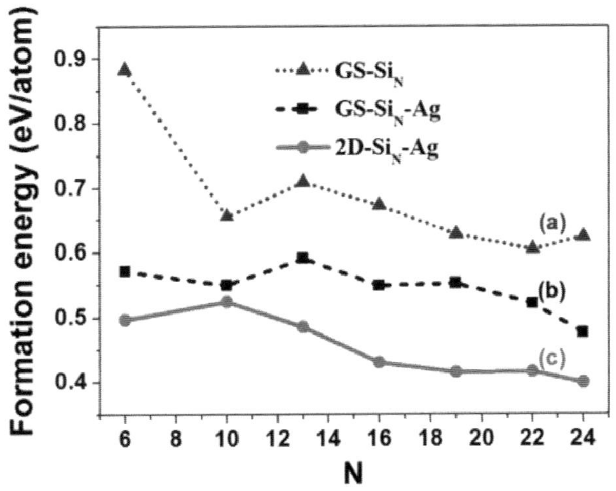

Figure 11. Formation energies of Si_N clusters in vacuum and on Ag(111) surface: (i) freestanding Si_N clusters with ground state configurations in vacuum (GS-Si_N); (ii) direct deposition of GS-Si_N clusters on Ag(111) surface (GS-Si_N-Ag); (iii) 2D silicene-like clusters on Ag(111) surface (2D-Si_N-Ag).[21]

In the fabrication of silicene or during the process of device manufacture, defects are almost inevitable. Here we briefly discuss the geometry and stability of some point defects in silicene monolayer, such as Stone-Wales rotation, single and double vacancies (abbreviated as SW, SV and DV, respectively, hereafter) and Si adatom.[31] The formation energy of a defect in silicene ε_F is defined as:

$$\varepsilon_F = \varepsilon_T - N \times \varepsilon_{Si} \qquad (2),$$

where ε_T is the total energy of defective silicene, N is the number of silicon atoms in the supercell of defective silicene, ε_{Si} is the energy per silicon atom in a perfect silicene sheet. There are two types of single and double vacancies as shown in Figure 12. The formation energy for a DV is small than two isolated SVs, implies that two SVs may coalesce into a DV by overcoming a migration barrier (0.12 eV for SV-1). The diffusion of a DV is much difficult. The barrier for creating a SW defect is 2.64 eV, while the reverse transition barrier is as low as 0.5 eV. Therefore, SW defects may be eliminated by annealing at appropriate temperature for a short time to achieve high-quality silicene. Surprisingly, adsorption of a Si adatom on silicene sheet is exothermic with negative formation energy of −0.03 eV. The Si adatom prefers the top site of silicene and the original lattice Si atom is pressed down (Figure 12f). Due to the negative formation energy, the influence of Si adatom needs to be carefully considered to produce high-quality silicene.

Moreover, the influence of electronic properties by defects must be taken into account. SW and DV-1(5|8|5) defects would introduce small band gaps of 33 meV and 161 meV in silicene respectively. SV-1(55|66) defect will transform semi-metallic silicene to metallic. DV-2(555|777) will destroy the Dirac cone, but keep the semimetal characteristic of silicene. Most interestingly, Si adatom not only opens a noticeable band gap of 91 meV, but also induces a magnetic moment of 2 μB in silicene.

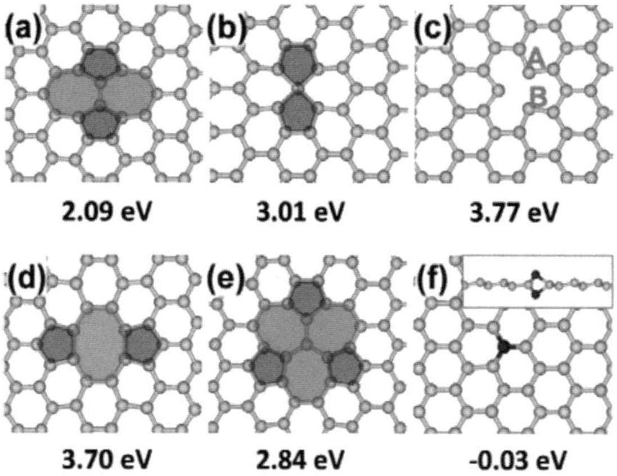

Figure 12. Geometries and formation energies of various point defects in silicene: (a) SW; (b) SV-1 by (55|66) rings; (c) SV-2 with three dangling atoms; (d) DV-1(5|8|5); (e) reconstructed DV-2(555|777); (f) Si adatom.[31]

4. Summary and prospect

In this minireview, substrate effects on silicene including insulating and metal substrate and the growth behavior of silicene on Ag were discussed. Some insulating substrates with appropriate work functions, such as *h*-BN monolayer and hydrogenated Si-terminated SiC(0001) surface, can preserve the Dirac cone of silicene. Thus, these substrates can provide effective supports for the applications of silicene in future microelectronic devices without destroying its geometry and electronic properties. Recently, there is an intense debate about whether Ag surface would preserve the linear dispersion of silicene, which is crucial for the possible application of silicene fabricated on metal substrates.

Moderate and homogeneous interaction between Si and Ag makes silicene very stable on Ag(111) surface even at a high temperature of 900 K. Thus, Ag(111) or other metal surface that has similar interaction with silicene may be good substrate for silicene growth. On Ag(111) surface, small 6MR-based silicene-like clusters are more stable than the ground state supercluster structures in vacuum. The reduction of formation energy with increasing cluster size and the easy diffusion of Si atom on Ag(111) surface ensure the facilitatory synthesis of silicene on Ag(111) surface. Some point defects may arise in silicene during the growth or fabrication process and greatly affect the electronic properties of silicene. Particular attention should be paid to Si adatom with negative formation energy and local magnetic moment.

At this moment, two major obstacles still obstruct in the middle of the road towards practical application of silicene in microelectronic devices. One is the competition and transformation between different silicene superstructures on Ag(111) surface. Deeply understanding the relationship between silicene reconstruction and substrate temperature is no doubt helpful to synthesize homogeneous large-scale silicene of high quality. Another important issue is to peel silicene from metal substrate. For microelectronic applications such as field effect transistors, silicene must be placed on insulating substrate. However, the epitaxial growth of silicene on insulating substrate has not realized yet. Therefore, how to peel silicene from metal surface and then transfer to inert substrate is vital for silicene application and is an urgent problem in silicene research. Alternatively, it is also desirable to directly synthesize silicene on insulating substrates.

Acknowledgment

This work was supported by the National Natural Science Foundation of China (No. 11304008), Development Fund of China Academy of Engineering Physics (No. 2013B0302056) and China Postdoctoral Science Foundation (No. 2013M530019).

References

[1] Takeda K and Shiraishi K. 1994 *Phys. Rev. B* **50**, 14916-14922.
[2] Cahangirov S, Topsakal M, Aktürk E, Şahin H and Ciraci S. 2009 *Phys. Rev. Lett.* **102**, 236804.
[3] Ding Y and Ni J. 2009 *Appl. Phys. Lett.* **95**, 083115.
[4] Pan L, Liu H J, Wen Y W, Tan X J, Lv H Y, Shi J and Tang X F. 2011 *Phys. Lett. A* **375**, 614-619.
[5] Şahin H, Cahangirov S, Topsakal M, Bekaroglu E, Akturk E, Senger R and Ciraci S. 2009 *Phys. Rev. B* **80**.
[6] Chen L, Liu C-C, Feng B, He X, Cheng P, Ding Z, Meng S, Yao Y and Wu K. 2012 *Phys. Rev. Lett.* **109**, 056804.
[7] Liu C-C, Jiang H and Yao Y. 2011 *Phys. Rev. B* **84**, 195430.
[8] Liu C-C, Feng W and Yao Y. 2011 *Phys. Rev. Lett.* **107**, 076802.
[9] Voon L C L Y, Sandberg E, Aga R S and Farajian A A. 2010 *Appl. Phys. Lett.* **97**, 163114.
[10] Osborn T H, Farajian A A, Pupysheva O V, Aga R S and Lew Yan Voon L C. 2011 *Chem. Phys. Lett.* **511**, 101-105.
[11] Houssa M, Scalise E, Sankaran K, Pourtois G, Afanas'ev V V and Stesmans A. 2011 *Appl. Phys. Lett.* **98**, 223107.
[12] Ding Y and Wang Y. 2012 *Appl. Phys. Lett.* **100**, 083102.
[13] Drummond N D, Zólyomi V and Fal'ko V I. 2012 *Phys. Rev. B* **85**, 075423.
[14] Ni Z, Liu Q, Tang K, Zheng J, Zhou J, Qin R, Gao Z, Yu D and Lu J. 2012 *Nano Lett.* **12**, 113-118.
[15] Lalmi B, Oughaddou H, Enriquez H, Kara A, Vizzini S, Ealet B and Aufray B. 2010 *Appl. Phys. Lett.* **97**, 223109.
[16] Vogt P, De Padova P, Quaresima C, Avila J, Frantzeskakis E, Asensio M C, Resta A, Ealet B and Le Lay G. 2012 *Phys. Rev. Lett.* **108**, 155501.
[17] Feng B, Ding Z, Meng S, Yao Y, He X, Cheng P, Chen L and Wu K. 2012 *Nano Lett.* **12**, 3507-3511.
[18] Lin C-L, Arafune R, Kawahara K, Tsukahara N, Minamitani E, Kim Y, Takagi N and Kawai M. 2012 *Appl. Phys. Express* **5**, 045802.
[19] Jamgotchian H, Colignon Y, Hamzaoui N, Ealet B, Hoarau J Y, Aufray B and Biberian J P. 2012 *J. Phys.: Condens. Matter* **24**, 172001.
[20] Chiappe D, Grazianetti C, Tallarida G, Fanciulli M and Molle A. 2012 *Adv. Mater.* **24**, 5088-5093.
[21] Gao J and Zhao J. 2012 *Sci Rep* **2**, 861.
[22] Fleurence A, Friedlein R, Ozaki T, Kawai H, Wang Y and Yamada-Takamura Y. 2012 *Phys. Rev. Lett.* **108**, 245501.
[23] Meng L *et al.* 2013 *Nano Lett.* **13**, 685-690.
[24] Houssa M, Pourtois G, Afanas'ev V V and Stesmans A. 2010 *Appl. Phys. Lett.* **97**, 112106.
[25] Liu H, Gao J and Zhao J. 2013 *J. Phys. Chem. C* **117**, 10353-10359.
[26] Guo Z-X, Furuya S, Iwata J-i and Oshiyama A. 2013 *J. Phys. Soc. Jpn.* **82**, 063714.
[27] Lin C-L, Arafune R, Kawahara K, Kanno M, Tsukahara N, Minamitani E, Kim Y, Kawai M and Takagi N. 2013 *Phys. Rev. Lett.* **110**, 076801.
[28] Wang Y-P and Cheng H-P. 2013 *Phys. Rev. B* **87**, 245430.
[29] Cahangirov S, Audiffred M, Tang P, Iacomino A, Duan W, Merino G and Rubio A. 2013 *Phys. Rev. B* **88**, 035432.
[30] Avila J *et al.* 2013 *J. Phys.: Condens. Matter* **25**, 262001.
[31] Gao J, Zhang J, Liu H, Zhang Q and Zhao J. 2013 *Nanoscale* **5**, 9785-9792.
[32] Enriquez H, Vizzini S, Kara A, Lalmi B and Oughaddou H. 2012 *J. Phys.: Condens. Matter* **24**, 314211.
[33] Resta A, Leoni T, Barth C, Ranguis A, Becker C, Bruhn T, Vogt P and Lay G. 2013 *Sci Rep* **3**, 2399.

[34] Bhattacharya A, Bhattacharya S and Das G. 2013 *Appl. Phys. Lett.* **103**, 123113.
[35] Novoselov K S, Geim A K, Morozov S V, Jiang D, Zhang Y, Dubonos S V, Grigorieva I V and Firsov A A. 2004 *Science* **306**, 666-669.
[36] Berger C *et al.* 2006 *Science* **312**, 1191-1196.
[37] Novoselov K S, Geim A K, Morozov S V, Jiang D, Katsnelson M I, Grigorieva I V, Dubonos S V and Firsov A A. 2005 *Nature* **438**, 197-200.
[38] Knox K R, Wang S, Morgante A, Cvetko D, Locatelli A, Mentes T O, Niño M A, Kim P and Osgood R M. 2008 *Phys. Rev. B* **78**, 201408.
[39] Castro Neto A H, Guinea F, Peres N M R, Novoselov K S and Geim A K. 2009 *Rev. Mod. Phys.* **81**, 109-162.
[40] Novoselov K S, Jiang D, Schedin F, Booth T J, Khotkevich V V, Morozov S V and Geim A K. 2005 *Proc. Natl. Acad. Sci. U. S. A.* **102**, 10451-10453.
[41] Pei S and Cheng H-M. 2012 *Carbon* **50**, 3210-3228.
[42] Léandri C, Oughaddou H, Aufray B, Gay J M, Le Lay G, Ranguis A and Garreau Y. 2007 *Surf. Sci.* **601**, 262-267.
[43] Aufray B, Kara A, Vizzini S b, Oughaddou H, Léandri C, Ealet B and Le Lay G. 2010 *Appl. Phys. Lett.* **96**, 183102.
[44] De Padova P *et al.* 2010 *Appl. Phys. Lett.* **96**, 261905.
[45] De Padova P, Quaresima C, Olivieri B, Perfetti P and Le Lay G. 2011 *J. Phys. D: Appl. Phys.* **44**, 312001.
[46] Vogt P, De Padova P, Quaresima C, Avila J, Frantzeskakis E, Asensio M C, Resta A, Ealet B and Le Lay G. 2012 *Physical Review Letters* **108**.
[47] Kara A, Enriquez H, Seitsonen A P, Lew Yan Voon L C, Vizzini S, Aufray B and Oughaddou H. 2012 *Surf. Sci. Rep.* **67**, 1-18.
[48] Ivanovskii A L. 2012 *Russ. Chem. Rev.* **81**, 571-605.
[49] Xu M, Liang T, Shi M and Chen H. 2013 *Chem. Rev.* **113**, 3766-3798.
[50] Yu-Ming L, Hsin-Ying C, Jenkins K A, Farmer D B, Avouris P and Valdes-Garcia A. 2010 *Electron Device Letters, IEEE* **31**, 68-70.
[51] Lemme M C, Echtermeyer T J, Baus M and Kurz H. 2007 *Electron Device Letters, IEEE* **28**, 282-284.
[52] Lin Y-M, Jenkins K A, Valdes-Garcia A, Small J P, Farmer D B and Avouris P. 2008 *Nano Lett.* **9**, 422-426.
[53] Farmer D B, Chiu H-Y, Lin Y-M, Jenkins K A, Xia F and Avouris P. 2009 *Nano Lett.* **9**, 4474-4478.
[54] Lin Y-M, Dimitrakopoulos C, Jenkins K A, Farmer D B, Chiu H-Y, Grill A and Avouris P. 2010 *Science* **327**, 662.
[55] Moon J S *et al.* 2009 *Electron Device Letters, IEEE* **30**, 650-652.
[56] Kedzierski J, Pei-Lan H, Healey P, Wyatt P W, Keast C L, Sprinkle M, Berger C and De Heer W A. 2008 *Electron Devices, IEEE Transactions on* **55**, 2078-2085.
[57] Gannett W, Regan W, Watanabe K, Taniguchi T, Crommie M F and Zettl A. 2011 *Appl. Phys. Lett.* **98**, 242105.
[58] Dean C R *et al.* 2010 *Nat. Nanotechnol.* **5**, 722-726.
[59] RadisavljevicB, RadenovicA, BrivioJ, GiacomettiV and KisA. 2011 *Nat. Nanotechnol.* **6**, 147-150.
[60] Ghatak S, Pal A N and Ghosh A. 2011 *ACS Nano* **5**, 7707-7712.
[61] Qiu H, Pan L, Yao Z, Li J, Shi Y and Wang X. 2012 *Appl. Phys. Lett.* **100**, 123104.
[62] Matsui T, Kambara H, Niimi Y, Tagami K, Tsukada M and Fukuyama H. 2005 *Phys. Rev. Lett.* **94**, 226403.
[63] Niimi Y, Kambara H, Matsui T, Yoshioka D and Fukuyama H. 2006 *Phys. Rev. Lett.* **97**, 236804.
[64] Li G and Andrei E Y. 2007 *Nat. Phys.* **3**, 623-627.

[65] Miller D L, Kubista K D, Rutter G M, Ruan M, de Heer W A, First P N and Stroscio J A. 2009 *Science* **324**, 924-927.
[66] Li G, Luican A and Andrei E Y. 2009 *Phys. Rev. Lett.* **102**, 176804.
[67] Gao J, Yip J, Zhao J, Yakobson B I and Ding F. 2011 *J. Am. Chem. Soc.* **133**, 5009-5015.
[68] Gao J, Yuan Q, Hu H, Zhao J and Ding F. 2011 *J. Phys. Chem. C* **115**, 17695-17703.
[69] Gao J and Zhao J. 2013 *Eur. Phys. J. D* **67**, 1-9.
[70] Zhu X and Zeng X C. 2003 *J. Chem. Phys.* **118**, 3558-3570.
[71] Zhu X L, Zeng X C, Lei Y A and Pan B. 2004 *J. Chem. Phys.* **120**, 8985-8995.
[72] Yoo S and Zeng X C. 2006 *J. Chem. Phys.* **124**, 054304-054306.

Thermal stability of standalone silicene sheet

Virgile Bocchetti[1,a], **H. T. Diep**[1,b,5], **Hanna Enriquez**[2,c], **Hamid Oughaddou**[2,3,d] **and Addelkader Kara**[4,e]

[1] Laboratoire de Physique Théorique et Modélisation, Université de Cergy-Pontoise, CNRS, UMR 8089, 2, Avenue Adolphe Chauvin, 95302 Cergy-Pontoise Cedex, France
[2] Institut des Sciences Moléculaires d'Orsay, ISMO-CNRS, Bâtiment 210, Université Paris-Sud, 91405 Orsay, France
[3] Département de Physique, Bâtiment Neuville II, 5 Mail Gay-Lussac, 95031 Neuville sur Oise, Cergy-Pontoise Cedex, France
[4] Department of Physics, University of Central Florida, Orlando, Florida, 32816 USA

E-mail: [a]virgile.bocchetti@u-cergy.fr, [b]diep@u-cergy.fr, [c]hanna.enriquez@u-psud.fr, [d]hamid.oughaddou@u-cergy.fr, [e]Abdelkader.Kara@ucf.edu

Abstract. Extensive Monte Carlo simulations are carried out to study thermal stability of an infinite standalone silicon sheet. We used the Tersoff potential that has been used with success for silicon at low temperatures. However, the melting temperature T_m calculated with the original parameters provided by Tersoff is too high with respect to the experimental one. Agrawal, Raff and Komanduri have proposed a modified set of parameters to reduce T_m. For comparison, we have used these two sets of parameters to study the stability and the melting of a standalone 2D sheet of silicon called "silicene", by analogy with graphene for the carbon sheet. We find that 2D crystalline silicene is stable up to a high temperature unlike in 2D systems with isotropic potentials such as Lennard-Jones. The differences in the obtained results using two sets of parameters are striking.

1. Introduction

Physics of 2D systems has been spectacularly developed during the last 30 years both experimentally and theoretically due to numerous applications in nanotechnology. In particular, we can mention the recent discovery of graphene with unusual properties such as high stability of 2D structure and remarkable transport behavior. For recent reviews given by experts on fundamentals and functionalities of graphene, the reader is referred to a special volume of MRS Bulletin [1]. Graphene is a Carbon sheet of one-atom layer thickness with large lateral size (up to a dozen of micrometers) discovered in 2004 [2, 3]. It has striking properties which can be applied in many domains such as optical materials, electronic materials, terahertz technology and spintronics. Intensive research activities are being carried out to incorporate graphene into devices for applications to exploit the novel properties of this unique nanomaterial (see recent reviews in Refs. [4, 5, 6, 7, 9, 8, 10] This is not an easy task. This difficulty leads scientists to turn to an alternative 2D version of silicon or germanium which would be easier to incorporate into the current silicon-based technology. In this paper we focus our attention in 2D silicon sheet known as silicene. Experiments have shown the existence of silicene nanoribbons [11, 12] on Ag(110) surface and a silicene sheet on Ag(111) surface [13, 14, 15, 16, 17]. An ordered two

[5] Corresponding author

dimensional surface alloy has been observed upon adsorption of silicon on Au(110) [18] and on a ZrB2 substrate [19]. Theoretical works using Molecular-Dynamics (MD) simulations [20] and density-functional theory (DFT) [21] have shown the stability of nanoribbons of different widths against temperature. Evidence of graphene-like electronic signature in silicene nanoribbons has been shown [22] and electronic structures of silicene fluoride and hydride have been revealed by calculations [23]. There is also a number of investigations on the stability of silicene sheet using a structure optimization [24], phonon dispersion and *ab initio* finite-temperature DFT theory. A low buckling of the silicene sheet has been suggested in these works. One of the problems in dealing with a silicon crystal concerns the choice of an appropriate interaction potential between Si atoms. In the bulk, a few potentials have been proposed to describe properties of bulk silicon crystals such as the Stillinger-Weber [25] and Tersoff potentials [26, 27]. These potentials have succeeded to reproduce principal physical properties of bulk Si crystal at low temperatures and to stabilize the diamond structure up to very high temperatures. However as seen below the melting temperature is largely over estimated with those potentials using MD simulations. To our knowledge, no Monte Carlo simulations on the subject have been reported, which triggered the interest in the subject..

In this paper, we show our results of constant-pressure MC simulations on the stability of a silicene sheet. In section 2 we present the Tersoff potential, our method and algorithm. In section 3 we present our results obtained with potentials using two different sets of parameters.

2. Model and Monte Carlo method

2.1. Tersoff potential

One of the most used potentials with the Metropolis MC algorithm is the Lennard-Jones potential (LJ) [28]. This potential cannot be employed to describe the silicene honeycomb structure. Such an isotropic potential always leads the system into the most compact face-centered-cubic (FCC) lattice when the temperature decreases. For Si crystals, the two potentials which are suitable at least at low temperatures are the Stillinger-Weber and Tersoff potentials. In this paper, we use the Tersoff potential. The reason of this choice will be explained below. This empirical potential was introduced by Tersoff in order to improve the accuracy in the description of properties of the Si crystal. Tersoff potential takes into account the dependence of the bond configuration on the local symmetry. This potential was introduced after the work of Ferrante, Smith, and Rose who have shown the universal behavior of calculated binding-energy curves for solid cohesion [29, 30]. We used here the Tersoff potential [26] given by the following expression:

$$V_{Tersof} = \frac{1}{2}\sum_{i\neq j} V_{ij} \tag{1}$$

where

$$V_{ij} = f_c(r_{ij})\left[A\exp\left(-\lambda r_{ij}\right) - b_{ij}B\exp\left(-\mu r_{ij}\right)\right] \tag{2}$$

$$b_{ij} = \chi\left(1+\beta^n\zeta_{ij}^n\right)^{\frac{-1}{2n}} \tag{3}$$

$$\zeta_{ij} = \sum_{k\neq i,j} f_c(r_{ik})g\left(\theta_{ijk}\right) \tag{4}$$

where

$$g\left(\theta_{ijk}\right) = 1+\frac{c^2}{d^2} - \frac{c^2}{\left[d^2+\left(h-\cos\theta_{ijk}\right)^2\right]} \tag{5}$$

and the cut-off function

$$f_c(r_{ij}) = \begin{Bmatrix} 1 & \text{if } r_{ij} < R, \\ \frac{1}{2} + \frac{1}{2}\cos\left[\pi\frac{(r_{ij}-R)}{S-R}\right] & \text{if } R < r_{ij} < S, \\ 0 & \text{if } r_{ij} > S \end{Bmatrix} \qquad (6)$$

In the above expressions, A, B, c, d, h, n, λ, μ, β, χ, R and S are constants. The above potential has two terms: the A term and the B term (see expression of V_{ij}). The A term is the two-body interaction while the B term incorporates many-body interactions including the three-body angles θ_{ijk}. This term determines the diamond structure of 3D silicon. Looking closely at the structure of the Tersoff potential, we realize that the two most important parameters are n which fixes the many-body interaction strength and h the parameter which determines the angle between two connected bonds. The original parameters given by Tersoff [26, 27] which yield a too high melting transition temperature ($T_m = 3600$ K) have been modified by Agrawal *et al.* to reduce it down to 2200 K which is closer to the experimental value $T_m(\exp)\simeq 1690$ K. These two different sets of parameters for the Tersoff potential are given in Table 1. Note that the main modifications concern precisely n and h, as discussed above.

Table 1. Tersoff parameters

Parameter	Orginal values (Ref. [26])	Values from ARK (Ref. [31])
A (eV)	1830.8	1830.8
B (eV)	471.18	471.18
λ (Å^{-1})	2.4799	2.4799
μ (Å^{-1})	1.7322	1.7322
β	1.1×10^{-6}	$\mathbf{1.15\times10^{-6}}$
n	0.78734	**0.988**
c	1.0039×10^5	1.0039×10^5
d	16.217	16.217
h	-0.59825	**-0.74525**
R (Å)	2.7	2.7
S (Å)	3.0	3.0
χ	1.0	1.0
T_m (Bulk Si)	3600 K	2200 K

2.2. The Monte Carlo algorithm

In our simulations, we consider a system of 968 atoms. The algorithm is split into two main parts: the construction of the honeycomb lattice with the minimization of the lattice energy at 0 Kelvin, and in the second part, the MC algorithm using the Metropolis updating criterion [32].

For testing purpose, we have build three different 2D planar lattices of silicon, namely honeycomb, square and triangular structures, and we have computed the energy per atom in this three different configurations. The results are shown in Fig. 1. As we can see, the honeycomb lattice is more stable than the two others at the nearest-neighbor (NN) distance $r = 2.31$ Åwhich agrees with experiments, using both the original set of parameters given by Tersoff [26] and the ARK parameters [31]: the energy has a minimum for the honeycomb lattice at that NN distance. For larger NN distances, the ARK parameters give an energy minimum for the square lattice at

$r = 2.42$ Å, but this distance does not correspond to the NN distance between Si atoms in the silicon crystal. So, at $T = 0$, we can say that both potentials give the same energy $E_0 \simeq 7.7$ eV and the same NN distance 2.31 Å.

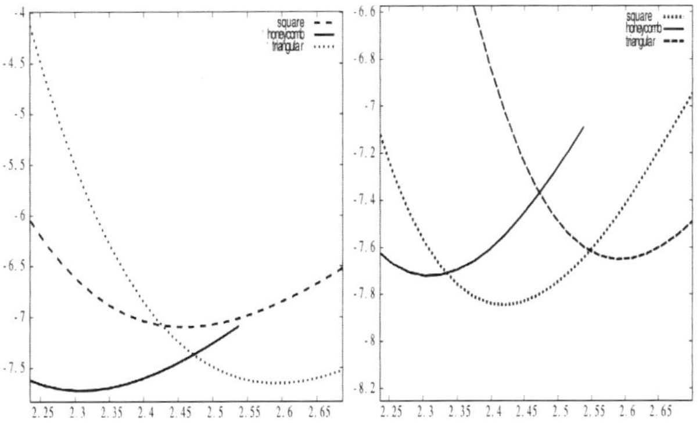

Figure 1. Energy per atom at 0 Kelvin,for three different lattices with the original set of parameters (left) and the ARK set of parameters (right). See text for comments.

Furthermore, calculations based on DFT [24] found that a buckling in the honeycomb structure of 0.4 Å stabilizes silicene. The buckled configuration is the stacking of two inverted triangular planes at a small z distance. In order to check this suggestion with our potential, we have introduced a parameter d which is the distance between two triangular planes and we have computed the energy per atom as a function of d. The potential used in the present study gives the planar structure ($d = 0$) as the energetically favorable structure. In order to reproduced the buckling found by the more accurate calculations, one need to re-fit all the parameters of the potential with the addition of one (such as d) ore more parameters; this is out of the scope of this study and will be tackled in futures studies.

For each MC step (MCS) we move all the atoms and we relax the size of the system. The magnitudes of the atomic displacements and the variation of the system size are determined so as to obtain an acceptation rate of about 50%. This collective updating is different from the single-atom updating algorithm which is not at all efficient for melting studies. Our algorithm allows a variation ratio of the volume; so the volume can make fluctuations (dilatation and contraction) around its equilibrium during the simulation time. An important fact is that the volume variation is controlled by the Metropolis algorithm, like the atom positions. For each simulation, we made approximately 3.10^7 MC steps per atom. Such long runs allow to observe the stability of the system and to overcome the very long relaxation time near the melting. At each MC step, after updating atom positions and relaxing the system volume, we compute the following transition probability

$$W = P\ (V_{new} - V_{old}) + (U_{new} - U_{old}) + N\ k_B\ T\ \ln\left(\frac{S_{new}}{S_{old}}\right) \tag{7}$$

where P is the pressure (0 in our case), S_{new} and S_{old} are respectively the new system surface and the old one, U_{new} denotes the new energy of the system after trial updating, U_{old} the old one, k_B the Boltzmann constant and T the temperature (in Kelvin).

A trial move is accepted if

$$\xi \leq e^{\left(\frac{-W}{k_B T}\right)} \tag{8}$$

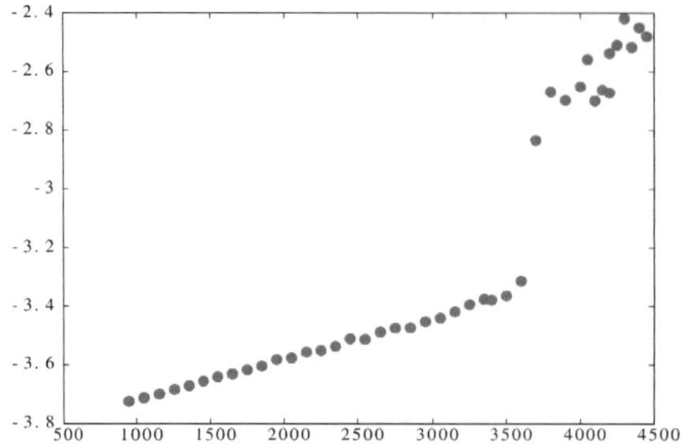

Figure 2. Energy versus temperature using original Tersoff parameters.

where ξ is a random number between 0 and 1. Otherwise, the system returns to its previous state with old atom positions and old surface size. We tune the magnitude of atom displacements and volume variations so as to have an acceptation rate of around 50%.

3. Results and discussion

For 2D systems with short-range isotropic interaction, it is known that long-range order does not survive at finite temperatures [33, 34]. Melting transition at finite T predicted by the Linderman criterion is for 3D crystals [35]. For the present 2D silicene, the potential is not isotropic because it stabilizes the Si diamond structure at very high T. As it turns out, this potential stabilizes also the honeycomb structure, as seen below. The stability of a silicene sheet can be observed by the energy versus temperature curve, the radial distribution function, snapshots of the system, the angular distribution function or the structure factor. We will show these quantities below.

In all our simulations, we started with a perfect lattice at 0 K and we increase the temperature to the interested temperature range.

In order to have more independent data and also to have faster computation, for each temperature we compute physical quantities of the system on an independent node of a CPU cluster. As an example, we show in Fig.2 the mean energy against temperature where each data point was computed by a node.

3.1. Results using original Tersoff parameters

At high T (see Fig. 2), we observe a first-order transition with a large latent heat. This transition is the melting of the sheet. Note that the melting temperature is very close to the simulated melting temperature of a 3D silicon crystal ($\simeq$ 3600 K) using the same potential. We calculate the structure factor $S_{\vec{K}}$ defined as

$$S_{\vec{K}} = \frac{1}{N_l} \left\langle \left| \sum_{j=1}^{N_l} e^{i\vec{K}.\vec{d_j}} \right| \right\rangle \tag{9}$$

where $\vec{d_j}$ is the position vector of an atom in the layer, N_l the number of atoms in a layer and $\vec{K}$ the reciprocal lattice vector which has the following coordinate (in reduced units): $2\pi\left(1; -\sqrt{3}; 0\right)$. The angular brackets $< ... >$ indicate thermal average taken over MC run time. The above "order parameter" allows us to monitor the long-range order. We show in Fig. 3 the structure factor versus T. As seen, the long-range order is lost at $T \simeq 3600$ K, namely at the temperature where the energy has a large discontinuity.

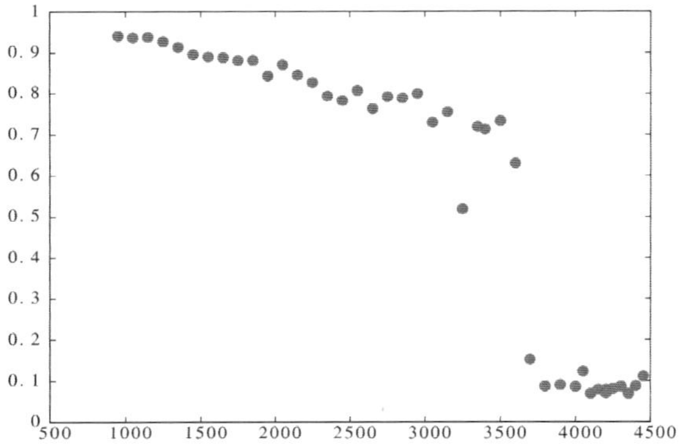

Figure 3. Structure factor versus temperature using original Tersoff parameters.

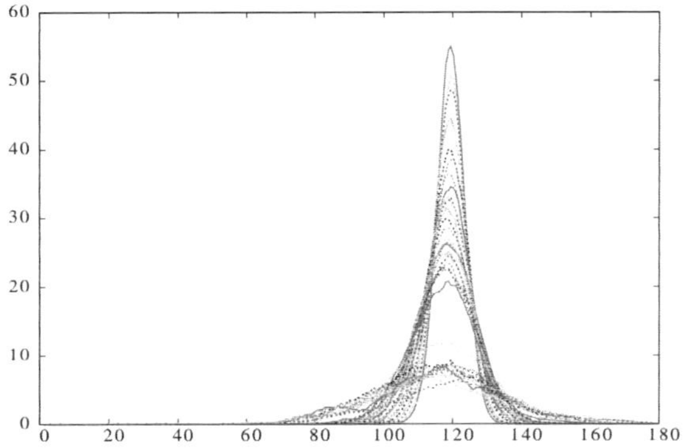

Figure 4. Angular distribution function for different T. See text for comments.

We show in Fig. 4 the angle distribution function at various temperatures. The pronounced peak at $T < 3500$ K undergoes a discontinuous decrease at $T \simeq 3600$ K. Furthermore, the angle distribution function does not show the appearance of any peak different from that at 120°. This shows the high stability of the honeycomb structure up to melting.

The radial distribution functions confirm the transition. When the temperature increases, the radial distribution function (Fig. 5) jumps from a state where we can distinguish the peaks corresponding to ordered positions of far neighbors to a state where only the peak of nearest-neighbors remains. The long-range order is lost.

In order to see how the number of nearest-neighbors evolves and indirectly how the density is modified, we have computed the integrated radial distribution shown in Fig. 5. As we can see, this number decreases from 3 in the perfect crystal to 2. The latter density corresponds to a wire structure of Si in 3D space (see Fig. 6). This behavior has been observed in the study of the melting of graphene [36].

3.2. ARK parameters

Let us recall that the experimental value of the bulk melting temperature T_m(exp) is about 1700 K. The Tersoff parameters yield T_m(Tersoff)$\simeq$ 3600 K while the ARK parameters used for the bulk Si crystal gives T_m(ARK)$\simeq$ 2200 K. So, the ARK parameters give a melting temperature closer to the experimental value.

In the case of a standalone sheet, the original Tersoff parameters, as shown above, give a very

Journal of Physics: Conference Series **491** (2014) 012008 doi:10.1088/1742-6596/491/1/012008

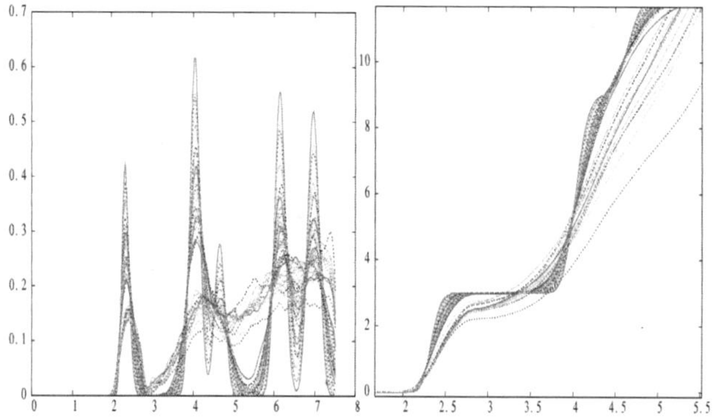

Figure 5. Radial distribution function for several T (left). Integrated radial distribution function for several T (right). See text for comments.

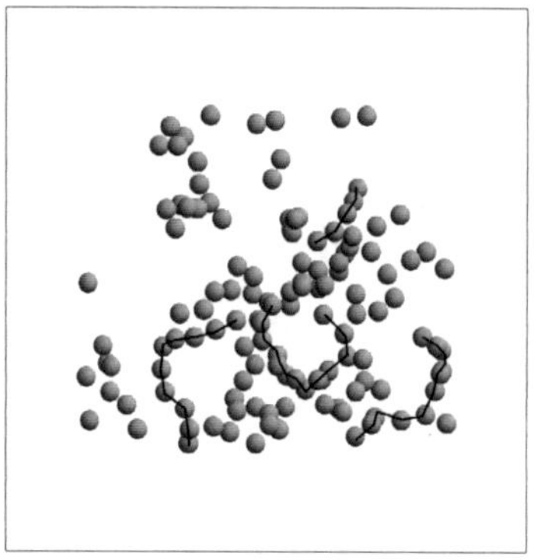

Figure 6. Picture of the system at high T (above the transition temperature). Disordered wire structures are observed.

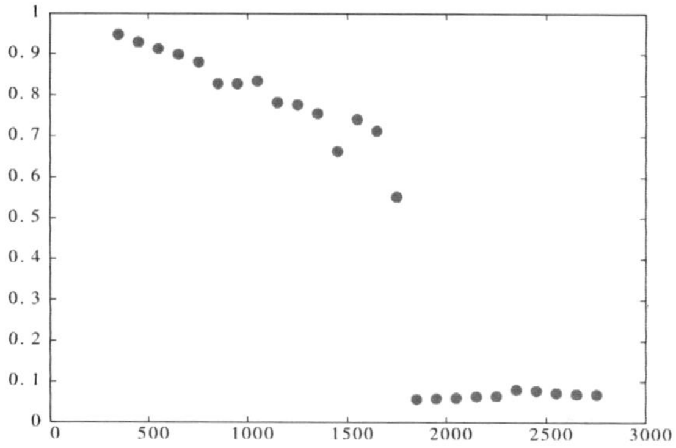

Figure 7. Structure factor versus temperature with ARK parameters.

high melting temperature, almost identical to that of the bulk, namely $T_m \simeq 3600$ K. Let us show now the melting temperature of a standalone sheet obtained by using the ARK parameters. We show the structure factor in Fig. 7. The long-range order is lost at $T_m \simeq 1750$ K, lower than

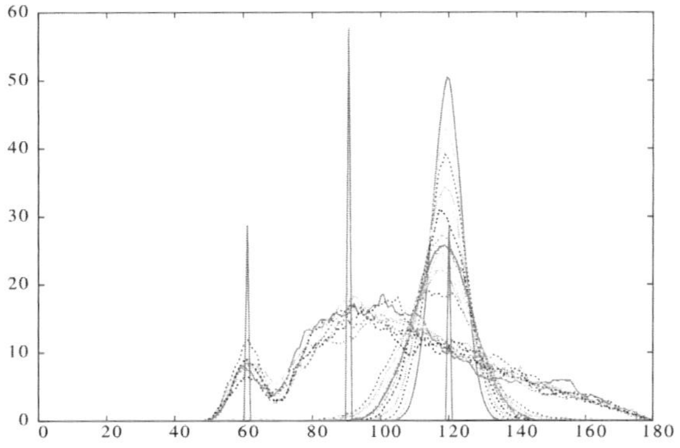

Figure 8. Angular distribution function for a wide range of T. Angular distribution function for the hexagonal 3D structure with three red peaks at 60°, 90° and 120° is shown for comparison.

that of the ARK bulk value, and less than a half of that obtained by using the original Tersoff parameters..

Furthermore, the angular distribution function (see Fig. 8) shows the apparition of two peaks at 60° and 90° at the same time a decrease of the peak at 120°. The honeycomb structure is thus strongly deformed to give rise to a 3D structure. For comparison, we show in Fig. 8 the angle distribution of the 3D hexagonal lattice. This result confirms the structural transition of the sheet to a 3D film of silicon in a structure where angles of 60° and 90° are proliferated.

In Fig. 9 we show the system at low and high T where we see the 2D structure before the transition and a 3D one at high T.

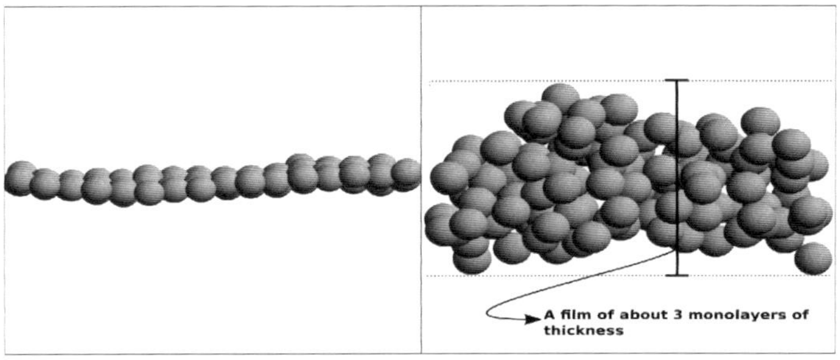

Figure 9. Instantaneous snapshots of the system shown by side views: the 2D structure below the transition (left) becomes a 3D one (right) above the transition.

The radial distribution and the integrated radial distribution shown in Fig. 10 indicate an abrupt transition with a jump of the number of nearest neighbors from 3 at low T to 4.6 at high T. This means that the silicon sheet is reorganized in a 3D structure which is more dense than the previous 2D one. The density jump in the integrated radial distribution is in agreement with the visual observation shown in Fig. 9.

4. Concluding remarks

In this paper, we studied the behavior of an infinite standalone silicene sheet. We have shown that the 2D honeycomb structure is stable up to high temperatures with the Tersoff potential. However, the temperature range of the silicene stability depends strongly on the parameters

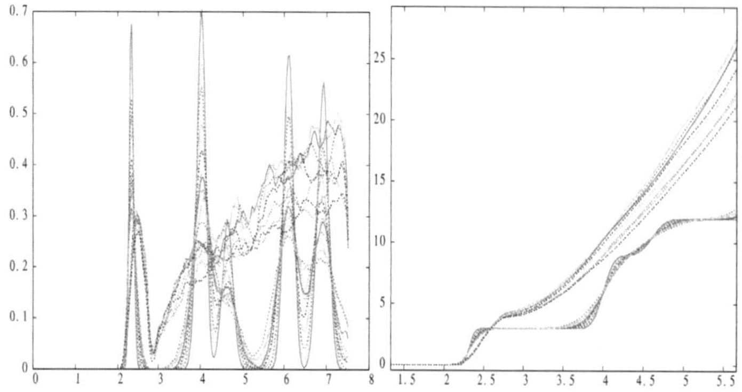

Figure 10. Radial distribution function for different T (left). Integrated radial distribution function for different T (right).

of that potential. The original Tersoff parameters give a too-high melting temperature while those proposed by Agrawal, Raff and Komanduri yield a melting temperature which is a half lower and much closer to 3D Si melting temperature. For both sets of parameters, the Tersoff potential gives rise to a silicene sheet without buckling. The flatness is stable with increasing temperature. Note that a very small buckling has been experimentally observed in the silicon deposition onto Ag(111) substrate maintained during the growth at 250 C [13] as well as in a theoretical DFT study [24]. There may be several reasons to explain the difference between these works and ours:

i) Bucking depends on the potential: for example with the Stillinger-Weber potential [25] we find that the ground state of silicene is buckled. However, with this potential, the flatness is progressively destroyed even far below the melting temperature. As said above, if we wish to introduce a buckling with the Tersoff potential, we should modify the whole set of parameters, not just add d to the potential.

ii) Buckling experimentally observed may come from the interaction with the substrate. Preliminary MC results from a Si ribbon deposited on Ag(111) show that the Si atoms do not remain on a flat plane. This is an interesting subject.

Finally, we note that the original Tersoff parameters make the system melt into a liquid of wires while the ARK ones make the system melt into a 3D uniform liquid. In view of the fact that the ARK parameters give a melting temperature closer to the experimental one, we believe that they also describe better the melting of the silicene sheet. The stability of a standalone sheet of silicene at high temperatures however is not yet tested in experiments in spite of the fact that it is experimentally proved that silicene on Ag (111) surface is stable at room temperature.

References

[1] *MRS Bulletin*, volume **37** (December 2012).
[2] Novoselov K S, Geim A K, Morozov S V, Jiang D, Zhang Y, Dubonos S V, Grigorieva I V, Firsov A A 2004 *Science* **306** 666
[3] Berger C, Song Z, Li T, Li X, Ogbazghi A Y, Feng R, Dai Z, Marchenkov A N, Conrad E H, First P N, de Heer W A 2004 *J. Phys. Chem.* B **108** 19912
[4] Singh V, Loung D, Zhai L, Das S, Khondaker S I, Seal S 2011 *Prog. Mater. Sci.* **56** 1178
[5] Haas J, de Heer W A, Conrad E H 2008 *J. Phys.: Condens. Matter* **20** 323202
[6] Allen M J, Tung V C, Kaner R B 2010 *Chem. Rev.* **110** 132
[7] Sarma S D, Adam S, Hwang E H, E. Rossi E 2011 *Rev. Mod. Phys.* **83** 407
[8] Araujo P T, Terrones M, Dresselhaus M S 2012 *Mater. Today* **15** 98

[9] Fuhrer M S, Lau C N, MacDonald A H 2010 *MRS Bull.* **35** 289
[10] Kuilla T, Bhadra S, Yao D, Kim N H, Bose S, Lee J H 2010 *Prog. Polym. Sci.* **35** 1350
[11] Kara A, Enriquez H, Seitsonen A P, Lew Yan Voon L C, Vizzini S, Aufray B, Oughaddou H 2012 *Surface Science Reports* **67** 118
[12] Aufray B, Kara A, Vizzini S, Oughaddou H, Léandri C, Ealet B, Le Lay G 2010 *Appl. Phys. Lett.* **96** 183102
[13] Lalmi B, Oughaddou H, Enriquez H, Kara A, Vizzini S, B. Ealet B, Aufray B 2010 *Appl. Phys. Lett.* **97** 223109
[14] Lin C L, Arafune R, Kawahara K, Tsukahara N, Minamitani E, Kim Y, Takagi N, Kawai M 2012 *Appl. Phys. Expr.***5** 045802
[15] Jamgoitcha H, Colignon Y, Hamzaoui N, Ealet B, Hoarau J Y, Aufray B, Bibérian J P 2012 *J. Phys.: Condensed Matter* **24** 172001
[16] Feng B, Ding Z, Meng S, Yao Y, He X, Cheng P, Chen L, Wu K 2012 *NanoLetters* **12** 3507
[17] Gao J, Zhao J 2012 *Scientific Reports* **2** 861
[18] Enriquez H, Mayne A, Kara A, Vizzini S, Roth S, Lalmi B, Seitsonen A P, Aufray B, Greber T, Belkhou R, Dujardin G, Hamid Oughaddou H 2012 *Appl. Phys. Lett.* **101** 021605
[19] Fleurence A, Friedlein R, Ozaki T, H, Wang Y, Yamada-Takamura Y 2012 *Phys. Rev. Lett.* **108** 245501
[20] Ince A, Erkoc S 2011 *Computational Materials Science* **50** 865
[21] Kara A, Vizzini S, Léandri C, Ealet B, Oughaddou H, Aufray B, Le Lay G 2010 *J. Phys.: Cond. Matter* **22** 045004
[22] Padova P D, Quaresima C, Ottaviani C, Sheverdyaeva P M, Moras P, Carbone C, Topwal D, Olivieri B, Kara A, Oughaddou H, Aufray B, Le Lay G 2010 *Appl. Phys. Lett.* **96** 261905
[23] Ding Y, Wang Y 2012 *Appl. Phys. Lett.* **100** 083102
[24] Cahangirov S, Topsakal M, Aktürk E, H. Åžahin H,Ciraci S 2009 *Phys. Rev. Lett.* **102** 236804
[25] Stillinger F H, Weber T A 1985 *Phys. Rev. B* **31** 5262
[26] Tersoff J 1989 *Phys. Rev. B* **39** 5566
[27] Tersoff J 1986 *Phys. Rev. Lett.* **56** 632
[28] Bernardes N 1958 *Phys. Rev.* **112** 1534
[29] Ferrante J, Smith J R, Rose J H 1983 *Phys. Rev. Lett.* **50** 1385
[30] Rose J H, Smith J R, Ferrante J 1983 *Phys. Rev.* **28** 1835
[31] Agrawal P, Raff L, Komanduri R 2005 *Phys. Rev.* B **72** 125206
[32] Allen M P, Tildesley D 1987 *Computer simulation of liquids* (London: Oxford Univ., Clarendon Press)
[33] Mermin D 1968 *Phys. Rev.* **176** 250
[34] Nelson D R, Halperin B I 1979 *Phys. Rev.* B **19** 2457
[35] Linderman F A 1910 *Phys. Z.* **11** 609
[36] Zakharchenko K V, Fasolino A, Los J H, Katsnelson M I 2011*J. Phys.: Cond. Matter* **23** 1

Chemical Synthesis of Silicon Nanosheets from layered calcium disilicide

Mustapha Ait Ali[1] **and Mohamed Rachid Tchalala**[1,2]

[1] Laboratoire de Chimie de Coordination et Catalyse, Département de Chimie, Faculté des Sciences-Semlalia, Université Cadi Ayyad, Marrakech, 40001, Morocco
[2] Institut des Sciences Moléculaires d'Orsay, ISMO-CNRS, Bâtiment 210, Université Paris-Sud, 91405 Orsay, France

Abstract. The synthesis of silicon nanosheets (SiNSs) without using conventional vacuum processes and vapor deposition is challenging and is anticipated to receive significant attention for a wide range of applications. Thereby, mass production of silicon nanosheets with low cost can be achieved using top-down approach starting from materials including a 2D sheet structure as a fundamental unit. We have recently, developed a soft synthetic method for the preparation of silicon sheets in gram-scale quantities by redox-assisted chemical exfoliation (RACE) of calcium disilicide ($CaSi_2$). In this review, we will present chemical methods for the synthesis and modification of silicon sheets.

1. Introduction

Low-dimensional nanomaterials are of particular interest as they may exhibit anisotropic and/or dimension-tunable properties, both of which are important attributes for nanodevice applications [1].

The silicon photovoltaic property discovered, two decades ago, by Canhamin in porous Si has not only challenged bulk semiconductor band structure models but also significantly expanded the potential applications of the Si-based materials [2,3]. The nanoscale properties of materials with nanometer dimensions differ from those of the bulk one. Nanostructured silicon materials compared to the bulk silicon materials exhibit photoelectronic effects, ballistic electron emission, much lower thermal conductivity and higher chemical activities. These properties lead to excellent potential application in photovoltaic cells, nonvolatile memory, lighting, ultrasonic generators, biosensors and other applications. Zero-dimensional (0D) crystalline Si, such as Si nanoparticles (SiNPs)[4,5] and one-dimensional (1D) nanomaterials, such as Si nanowires SiNWs [6,7] and Si nanotubes (SiNTs)[8,9] have been synthesized. The fabrication techniques have been intensively studied to improve size and chemical stability, as well as to achieve high yield. Silicon Nano-sheets (SiNSs), a low dimensional crystalline silicon materials with their high specific surface area, makes them promising candidates for a variety of applications in nanoscience and nanotechnology. There are two prominent types of SiNSs:

- buckled sheet mainly based on the Si (111) structure, with the same honeycomb lattice structure, but is corrugated due to sp^3 hybridization.
- silicene with a graphene-like honeycomb lattice structure and a mixed sp^3-sp^2 hybridization, that is weakly corrugated [10].

Many types of nanosheets that consist of oxidized substrates, such as silicate layers or metal oxides, have been reported.[11-15] However, buckled Si sheets or silicene have not been extensively developed.

Before addressing chemical synthesis methods of silicon sheets, a comparison of chemical and physical properties of silicon with to those of carbon are first briefly outlined in table 1. More details can be found in several reviews and books [16-18].

Silicon and Carbon are in Group 14 of the periodic table. They have four electrons in the outermost shell, four valence orbitals witch lead to formation of four two-electron covalent bonds. However, there are many differences between these two elements in terms of their structural, chemical and physical properties. These differences affect the ability for formation of their respective nanomaterials and nanostructures.

Table 1. Silicon vs Carbon: Similarities and Differences

Carbon	Silicon
Group 14 in the periodic table	Group 14 in the periodic table
four electrons in the outermost shell four valence orbitals formation of four two-electron covalent bonds	four electrons in the outermost shell four valence orbitals formation of four two-electron covalent bonds
Paulings electronegativity χ 2.55 for C and 2.20 for H	lower electronegativity $\chi = 1.90$
polarization of the C-H bond $C^{\delta-} - H^{\delta+}$	opposite sense polarization $Si^{\delta+} - H^{\delta-}$
C-H bonds are rather strong	Si-H more reactive
the valence s and p orbitals with large energy difference (10.60 eV) giving rise to sp, sp^2, and sp^3 hybridizations	smaller energy difference between the s and p orbitals (5.66 eV), lower hybridization energies. Silicon tends to use all its three p orbitals, resulting in sp^3 hybridization
no energetically accessible d orbitals to expand its coordination or valence shell	d orbitals (or other virtual orbitals) are energetically low-lying allowing silicon to expand its coordination sphere from four to five or six.
small atomic size means efficient $\pi-\pi$ overlap and the formation of stable double and triple bonds C-C, C=C, and C≡C bond lengths are 1.54, 1.34, and 1.20Å	larger atomic size of silicon results in poor $\pi - \pi$ overlaps Si-Si and Si=Si bonds length of 2.35 and 2.16Ålong.
C=C 1.54Åresult in planar structures	Si=Si,2.16Åinstable with nonplanar structures, often folded and sometimes twisted, or both.

Stable crystalline silicon typically assumes only the sp^3 hybridization system and favors a tetrahedral diamond-like structure. Silicon equivalent of graphene was first mentioned in a theoretical study by Takeda and Shiraishi in 1994 [19] and then reinvestigated by other groups, who named it silicene [20-23]. Silicene with a honeycomb graphene-like structure has been observed recently by scanning tunneling microscopy (STM) on Ag(110) and on Ag(111) surfaces [24,25,26]. Silicene formation has also been studied on Au(110) [27], Ir(111) [28] and on ZrB_2 [29] surfaces . So far, these silicon structures have been grown using epitaxial growth techniques under UHV. However, silicene does not seem to exist in nature and there is no solid phase silicon similar to graphite. This is inconsistent with theoretical studies that have predicted the stability of silicene nanosheets inserted into a graphite-like lattice, such as ultrathin AlN stacks [30] or modified lattice structures through incorporation of transition-metal ions(M=Ti, Nb, Ta, Cr, Mo, and W) to form MSi_2 [31]. Zintl phases metal disilicides that incorporate rare-earth metals and alkaline earth metals, MSi_2 (M=Yb, Er,Tm, Lu and Ca), have graphite-like, stacked 2D silicon sub-network structures that are isomorphous with AlB_2 [32]. Such binary compounds are synthesized by direct reaction between the metal and silicon in a helium atmosphere at $T > 900°C$ [33]. Topochemical transformation of layered Zintl phases can be used to obtain sheet polymers. Nevertheless, the exact stoichiometry of the polymers obtained and their properties depend on the specific reactions taking place.

Recently, we have developed a soft synthetic method for silicon sheets, with hexagonal graphitic-like structure resembling that of silicene, in gram-scale quantities by redox-assisted chemical exfoliation (RACE) of calcium disilicide ($CaSi_2$) [34].

This review focuses on chemical synthesis and modification of silicon sheets. We shall begin by topochemical reactions of layered Zintl phases in section 2 with the well-known reactions of $CaSi_2$ investigated more recently. A brief structural discussion of calcium disilicide ($CaSi_2$) that includes 2D silicon puckered will be introduced. The Preparation of SiNSs by surface modification of Si_6H_6 will be presented in section 3. Finally, our conclusions and future plans for silicon sheet will be presented in section 4.

2. Synthesis of SiNSs by topochemical reactions of the layered Zintl phases

The top-down solution approach is recently a reliable way to synthesize SiNSs by using materials including a 2D sheet structure as a fundamental unit.

Alkali and alkaline earth silicides belong to the family of Zintl phases, in which polyanionic species, formed of the tetrelide [35] atoms counter balance the electropositive metal cations [36-38]. These compounds are interesting because the Si atom occur with negative oxidation states and low coordination numbers, and they form unusual low-dimensional structures such as sheets, chains or isolated polyhedra. They are also important precursors for the synthesis of bulk amorphous semiconductors, including Si nanoparticles prepared by solution- phase routes or in the solid state [39] and also the expanded framework solids known as semiconductor clathrates that are isomorphous with ice clathrate hydrates [40-42]. Among the Zintl silicides, only calcium disilicide ($CaSi_2$) has a structure of interconnected Si_6 rings forming anionic 2D silicon puckered sheets $(Si_2)^{2-}_{\infty}$ held together by Ca^{2+} that form planar monolayers (Figure 1). The covalent structure of the silicide layers is stable and remarkably flexible [43-45] and can be chemically modified by oxidation, leading to removal of the intercalated metal ions.

2.1. $CaSi_2$ as precursor material for the preparation of silicone nanosheets

Removal of Ca^{2+} from $CaSi_2$ has been achieved by Bonitz et al. [46] during treatment of a suspension of $CaSi_2$ in CCl_4 with gaseous Cl_2 (eq. 1).

$$(CaSi_2)_n + nCl_2 \rightarrow (Si)_{2n} + nCaCl_2 \quad (1)$$

The 2D silicon sheets obtained according to equation (1) is unstable, highly reactive, and cannot be easily handled and characterized.[47, 48]

In 1863, Whler reported for the first time the synthesis of stable siloxene sheet polymer ($Si_6H_3(OH)_3$) by topochemical transformation of $CaSi_2$ in ice-cold aqueous HCl solution as a

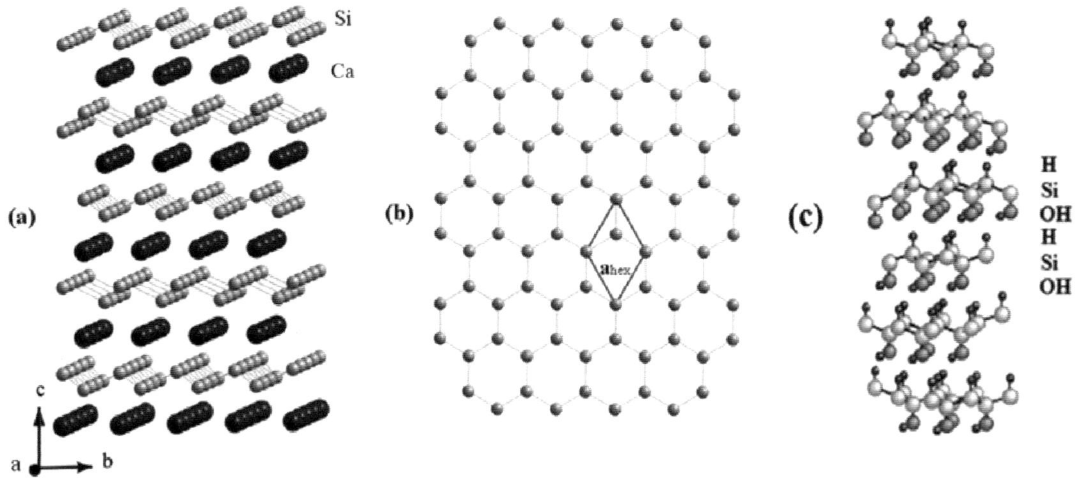

Figure 1. (a) Crystal structure of $CaSi_2$ (cell parameters: a=3.855 Åand c=30.60Åwith space group $R-3m$). The large circles represent Ca atoms and the small circles represent Si atoms. (b) Silicon sheet in $CaSi_2$ crystal structure.(c) Weiss siloxene $Si_6H_3(OH)_3$. All OH groups are replaced to H atoms in layered polysilane Si_6H_6.

yellow compound called "Silicon" (eq. 2).[49] During this transformation, Ca^{2+} ions are removed without destroying the polyanion layers, which then forms the backbone of the resulting sheet polymer.

$$3CaSi_2 + 6HCl + 3H_2O \rightarrow Si_6H_3OH_3 + 3CaCl_2 + 3H_2 \quad (2)$$

Kautsky later modified the reaction conditions and obtained a gray-green amorphous product "siloxene" [50]. This material is of considerable interest due to its intense luminescence and the possibility of shifting the luminescence wavelength through substitution [51, 52].

Weiss et al. in 1979 [45] proposed that the crystalline sheet polymer structure was a puckered Si 2D layers similar to crystalline Si(111) layers stabilized though termination of hydrogen or hydroxide groups pointing out of the layer plane. This suggestion was later confirmed by Dahn et al.(Figure 1).[53] The use of a solution of HCl in methanol or ethanol instead of an aqueous solution led to the preparation of alkoxide-terminated siloxenes ($Si_6H_3(OCH_3)_3$ or $Si_6H_3(OC_2H_5)_3$) (eq. 3). [54]

$$3CaSi_2 + 6HCl + 3ROH \rightarrow Si_6H_3(RO)_3 + 3CaCl_2 + 3H_2 \quad (3)$$

R= methyle, ethyle

The treatment at temperatures below -30°C resulted in non-oxidized silicon sheet polymer; a layered polysilane (Si_6 H_6), with similar structure to $Si_6H_3OH_3$, but in which Si atoms are terminated by hydrogen (Figure 1) (eq. 4) [55, 56].

$$3CaSi_2 + 6HCl \rightarrow Si_6H_6 + 3CaCl_2 \quad (4)$$

Performing the reaction at 0°C, Dahn et al. obtained the same layered polysilane Si_6 H_6 [53]. However,Yamanaka et al. reported formation of $Si_6H_3OH_3$ as by-product which was then removed by washing with a solution of HF [55].

During topochemical transformation, $CaSi_2$ was spontaneously cleaved to yield siloxene platelets by reactive deposition epitaxy (RDE) method.[49]

Layered siloxene $Si_6H_3OH_3$ was successfully exfoliated into individual nanosheets with a thickness of 0.7 nm and lengths in the range 100-200 nm by Nakano et al. [57], using sodium dodecyl sulfate as dispersant agent (Figure 2). Sodium dodecylsulfate (SDS) and thin plate-like crystals of $Si_6H_3OH_3$ were shaken at room temperature for 10 days, to give a translucent colloidal suspension. The suspension remained stable for two months without precipitation. This can be used to synthesize $Si_6H_3OH_3$nanosheets, but the silicon skeleton is partially oxidized.

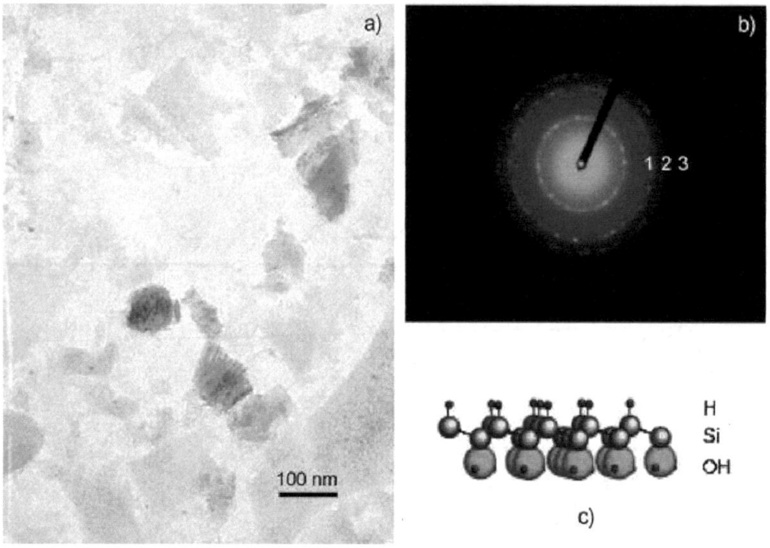

Figure 2. a) TEM image of $Si_6H_3(OH)_3$ nanosheets. b) Electron diffraction pattern for the $Si_6H_3(OH)_3$. Indices and spacings for the diffraction rings 10 and 0.313 nm (ring 1), 11 and 0.188 nm (ring 2), and 20 and 0.163 nm (ring 3), respectively. c) Schematic illustration of the $Si_6H_3(OH)_3$nanosheet. Reproduced from reference [57].

2.2. Synthesis from Mg-doped $CaSi_2$ ($CaSi_{1.85}Mg_{0.15}$)

One of the most important techniques to exfoliate single layers from their parent layered compound is the adjustment of the charge on the silicon layer in $CaSi_2$.To weaken the electrostatic interaction between the Ca^{2+} and $((Si_2)^{2-}_{\infty}$ layers, in $CaSi_2$, it is important to reduce the charge on the negatively charged silicon layers. For that, Nakano et al. [58] have chosen Magnesium $^{(Mg2+)}$ as electro-positive element to reduce silicon charge in $CaSi_2$. Mg-doped $CaSi_2$, ($CaSi_{1.85}Mg_{0.15}$) was prepared and used as a precursor for the chemical synthesis of silicon sheets. They prepared $CaSi_{1.85}Mg_{0.15}$ by melting a mixture of CaSi, Si, and Mg ; then the material obtained was treated in a solution of propylamine hydrochloride (PAHCl), resulting in the de-intercalation of Ca^{2+} ions and the formation of a light-brown suspension containing silicon nanosheets (SiNSs). The following mechanism for the exfoliation process was proposed: 1) oxidation of $CaSi_{1.85}Mg_{0.15}$ with PAHCl, accompanied by the liberation of PA; 2) the presumably very reactive Mg-doped Si_6H_6 obtained is readily oxidized by water to form gaseous hydrogen; and 3) Mg-doped layered silicon with capping oxygen atoms is exfoliated by reaction with the aqueous solution of PA.

TEM image (Figure 3) revealed a 2D structure with lateral dimensions in the range of 200 to 500 nm. The composition of the obtained monolayer SiNSs was determined to be

Si/Mg/O=7.0:1.3:7.5 by X-ray photoelectron spectroscopy (XPS).Monolayer and ultrathin sheets were evidenced by Atomic force microscopy (AFM) (Figure 3) with an average thickness of 0.37 nm measured at intervals between the sheets and the substrate surface.

The Electron Diffraction pattern (Figure 3) confirmed that the sheet was single crystalline and the diffraction spots could be indexed as hk reflections of a hexagonal lattice with a=0.82 nm, which corresponds to approximately twice of the interplanar distance of the (111) planes of bulk silicon (0.38 nm).Therefore, the exfoliation of Mg-doped $CaSi_2$ using PAHCl produces less than 1% of surface-oxidized silicon nanosheets (SiNSs). This oxygen capping is inherent to the process, because the capping is thought to promote exfoliation when the layered silicon becomes a stable colloidal suspension by reaction with aqueous propylamine solution. Despite these studies, the chemical synthesis of pure silicon sheets has not yet been reported.

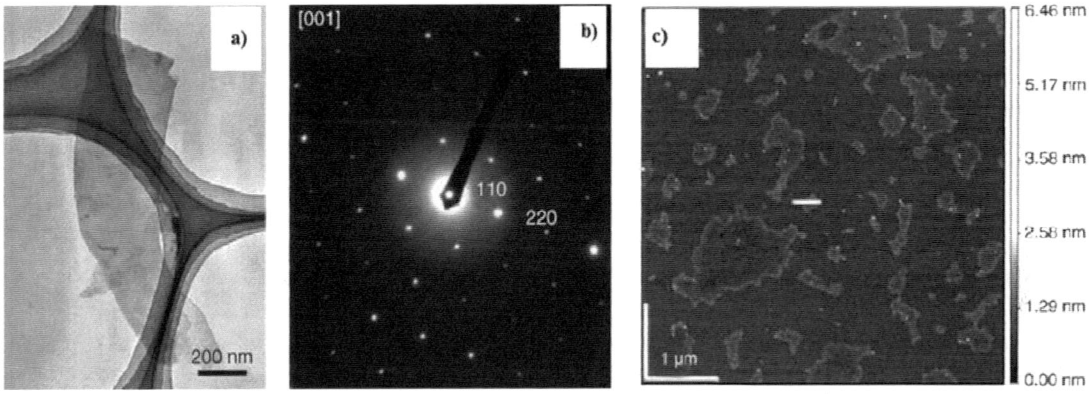

Figure 3. a) TEM image of the sheet. b) ED pattern recorded along the [001] zone axis perpendicular to the surface of the sheet. C) Noncontact mode AFM image of the silicon sheets. Reproduced from reference [58]

2.3. Redox assisted chemical exfoliation of $CaSi_2$

Direct chemical synthesis of pure silicon sheets in gram-scale quantities has been recently reported by our group using redox-assisted chemical exfoliation (RACE) of potassium modified calcium disilicide ($CaSi_2$).[34] Let us recall that $CaSi_2$ is an ionic ($Ca^{2+}(Si_2)^{2}-$) material, and the reduction of the charge on the (negatively charged) silicon layers is an important step in order to weaken the electrostatic interaction between the Ca^{2+} and $(Si_2)^{2-}_{\infty}$ layers.[58] For that, we have chosen potassium (K) as doping element with low melting point (63.2C), high reactivity and lower electronegativity. We proposed the following mechanism for the exfoliation process:

(1) Diffusion of potassium (K) in $CaSi_2$ at 160 °C, leads to formation of the $K_{2x}Ca_{1-x}Si_{2-x}$ phase and a small amount of KSi and Ca metal driven by the reduction of some Ca^{2+} by K, as shown in the following chemical equation (5) :

$$CaSi_2 + 3xK \rightarrow K_{2x}Ca_{1-x}Si_{2-x} + xKSi + xCa \tag{5}$$

(2) The obtained mixture was then treated with a degassed solution of isopropylamine hydrochloride (i-PAHCl) in ethanol at room temperature for 24 hours under nitrogen. The calcium ions are deintercalated accompanied by the evolution of hydrogen (eq. (6)) and a greenish-brown suspension was formed accompanied with an insoluble black solid which was assigned to the unreacted $CaSi_2$ after confirmation by XRD study. The SiNSs was obtained

after filtration of the greenish-brown suspension in 20% yield. These steps can be summarized by the following three chemical equations:

$$K_{2x}Ca_{1-x}Si_{2-x} + 2iPA.HCl \rightarrow 2xKCl + (1-x)CaCl_2 + (2-x)Si(SNSs) + 2iPA + H_2 \quad (6)$$

It is important to note that when adding the iPA.HCl solution, a strong evolution of hydrogen is observed which may be due to excess potassium oxidation (eq. (8)). The rate of hydrogen evolution involving the oxidation of deintercalated calcium (eq. (7)) accompanied by liberation of nanosheets of silicon, decreases significantly after a few minutes.

$$Ca + 2iPA.HCl \rightarrow CaCl_2 + 2iPA + H_2 \quad (7)$$

The excess potassium reacts with iPA.HCl to release H_2 as follows:

$$K + iPA.HCl \rightarrow KCl + 2iPA + 1/2H_2 \quad (8)$$

The synthesized sheets were analyzed by X-ray photoemission spectroscopy (XPS), high- resolution transmission electron microscopy (HR-TEM) and energy dispersive X-ray spectroscopy (EDX) (Figure 4). XPS measurements show that the synthesized sheets contain almost exclusively Si atoms and displayed a Si 2p peak with a binding energy of 100 eV, which is different from that of SiO_2 (104 eV) (Figure 4a). Energy-dispersive X-ray spectroscopy (EDX) experiments show a small trace of oxygen which is due to exposure of the powder to air. HR-TEM experiments (Figure 4b) reveal that the synthesized silicon product consists of highly ordered stacked silicon sheets. Electron diffraction pattern (Figure 4d) shows hexagonal arrangement, consistent with well-ordered and crystalline silicon sheets.

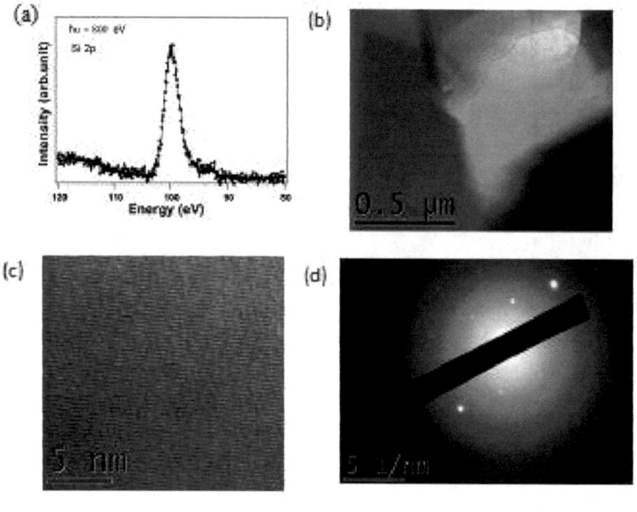

Figure 4. a) XPS spectrum of the synthesized Si sheets, the Si 2p core levels shown, b) TEM image showing a stack of silicon sheets, c) TEM image showing a highly oriented crystalline silicon sheet, d) ED pattern recorded perpendicular to the surface of the sheet. Reproduced from reference [35]

This hexagonal pattern originates either from a quasi-two dimensional hexagonal lattice, with the observed spots corresponding to the four (100) and the two (110) reflections. From the interplanar distance d_{100} (or d_{110}), the 2D hexagonal lattice constant a_{hex}, was calculated to be $(0.330 \pm 0.018) nm$ which is smaller than the 0.38 nm corresponding to the (111) plane structure of bulk silicon. Therefore, these silicon sheets may correspond to a 2D honeycomb Si lattice stacked in planes with a surface lattice constant $a_{hex} = (0.330 \pm 0.018) nm$, corresponding to a distance between the first Si atoms neighbors of 0.19 nm. Previous studies reported that silicene is corrugated [25,28,10,59-61] with a buckling between 0.04 and 0.12 nm [62,63], resulting in a projected distance (in the plane of the hexagons) as low as 0.19 nm. The data clearly show that the synthesized Si sheets do not correspond to the (111) layer of Si diamond lattice; instead they have a hexagonal graphitic-like structure resembling that of silicene.

3. Preparation of SiNSs by surface modification of Si_6H_6

Capping of layered silicon compounds with organic groups rather than oxygen atoms is supposed to promote their exfoliation, and assist in the control of the interlayer spacing. Moreover, the synthesis of organically modified layered silicon allows for the preparation of carbon-coated silicon nanosheets by mild non-oxidative pyrolysis.

The introduction of organic groups on Si_6H_6 through $Si - C$ or $Si - N$ bonds is plausible method to obtain SiNSs capped with organic moieties, because of their high dispersion in organic solvents and their stability with respect to oxidation and/or hydrolysis, due to formation of hydrophobic surface on silicon. In contrast to carbon, silicon ($\chi = 1.90$) is less electronegative than hydrogen ($\chi = 2.20$); hence, Si-H bonds are polarized in the opposite sense, as $Si^{\delta+}$-$H^{\delta-}$. This implies that nucleophilic attacks on silanes usually occur at the silicon centers. Silanes are also much more reactive than the corresponding hydrocarbon analogues due to larger polarization. The high reactivity of silanes and hydride-terminated silicon crystalline surfaces are also due to weak $Si - H$ bonds, which are significantly weaker than $C - H$ bonds.

Based on this characteristic, layered polysilane, Si_6H_6, is a suitable starting material for the preparation of non-oxidative silicon nanosheets by organic modification and exfoliation. Si_6H_6 can be prepared from calcium disilicide ($CaSi_2$),[55,56] and each layer of Si_6H_6 consists of two-dimensional corrugated Si(111) planes, in which the Si_6 rings are interconnected and a H atom is bound to each Si atom in the direction perpendicular to the layer.[45,53,55]

3.1. Hydrosilylation reaction

Surface modification of silicon particles or substrates with organic groups, can be achieved through hydrosilylation of hydrogen-terminated silicon surface with unsaturated organic groups, such as alkenes and alkynes.[64a] However, this reaction did not proceed readily for Si_6H_6. Si-H bonds as reactive sites are not readily accessible because they are located in the interlayer spacing of the layered starting material. Therefore for the reaction to happen, it is imperative that the organic moiety must be intercalated into the interlayer. The reactivity of Si_6H_6 must be increased by halogenations;[64b] however, this is undesirable, because highly toxic gases such as PCl_5, CCl_4, or Cl_2 are required.

Recently, Nakano et al. [65] have functionalized Si_6H_6 with hexyl chains using a Pt-catalyzed hydrosilylation reaction with 1-hexene to produce a stable colloidal suspension (Scheme 1). The obtained product was soluble in typical organic solvents, such as hexane, chloroform, acetone, and ether, and insoluble in water and ethanol.

FTIR and Si K-edge X-ray absorption near-edge structure (XANES) analyses (Figure 5) revealed that the 1-hexene does not merely attach to Si_6H_6, but reacts with it to form Si-C linkages. The IR spectrum (Figure 5a) shows an alkyl stretching and bending absorptions bands between 2856 and 2954 cm^{-1} and between 1259 and 1459 cm^{-1}, respectively. In addition, a band attributable to $Si - CH_2$ vibrational scissoring was observed at 729 cm^{-1}. The presence of characteristic vibrations for organic molecules in the sheets after the reaction, along with a reduction in the intensity of $Si - H$ at 2110 cm^{-1} and the absence of bands characteristic of a

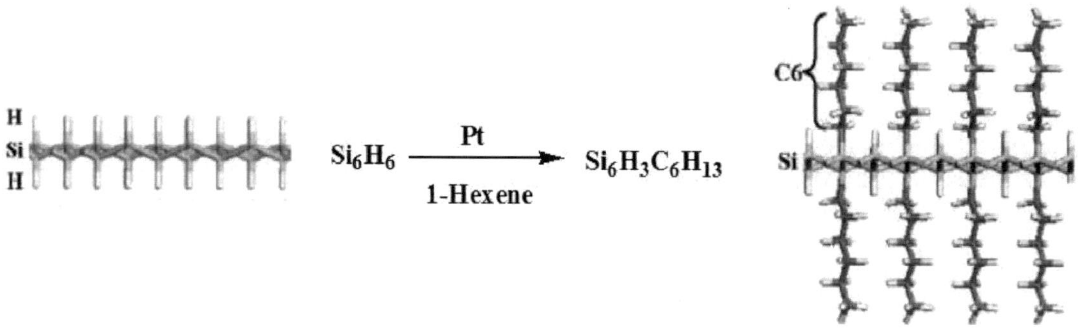

Scheme 1: Hydrosilylation of Si_6H_6 with 1-hexene catalyzed by platinum

terminal double bond ($C = C$; 1600 cm^{-1}), indicate that the organic molecules were covalently attached to silicon surface. However, $Si - O$ stretching vibrations bands observed between 1000 and 1100 cm^{-1} indicate that the synthesized sheets were partially oxidized. XANES of prepared silicon sheet (Figure 5b) exhibits features at an energy of 1844 eV, which is intermediate between those measured for Si and SiO_2 samples. This peak was assigned to the $Si - C$ bond, which is the same bond energy as that for a tetra-coordinated silicon atom-bonded organic group.[66a]

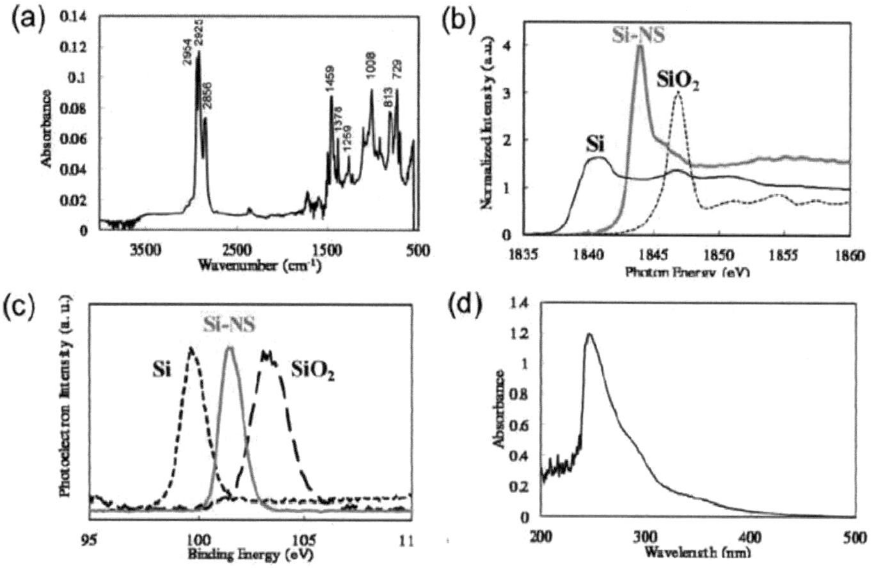

Figure 5. (a) FTIR spectra of $[Si_6H_3(C_6H_{13})_3]$. (b) Si K-edge XANES spectra of $[Si_6H_3(C_6H_{13})_3]$ (brown line), Si(111) wafer and SiOx (black line). The data was recorded at BL-10 in the Ritsumeikan SR Center. (c) XPS Si 2p spectra of $[Si_6H_3(C_6H_{13})_3]$ (brown line), Si(111) wafer and SiOx (black line). (d) UV spectrum of $[Si_6H_3(C_6H_{13})_3]$. Reproduced from reference [65]

The XPS spectrum displayed a Si 2p peak with a binding energy of 102.0 eV (Figure 5c), which is different from those of bulk Si (99.0 eV) and SiO_2 (104 eV). Compared to XPS data for decyl-capped silicon nanocrystals reported recently,[67] this binding energy (102.0 eV) was

Journal of Physics: Conference Series **491** (2014) 012009 doi:10.1088/1742-6596/491/1/012009

attributed to a Si-organo group. AFM of the sheet gave a thickness of 3.1 nm (Figure 6), which was compared to thickness of 2.3 nm calculated from model structure optimized using force field mode in Discover program.[65] The model structure was constructed on the basis of the ideal formula [$Si_6H_3(C_6H_{13})_3$] (Scheme 1). The thickness indicated that the sample was composed of monolayer sheets [65].

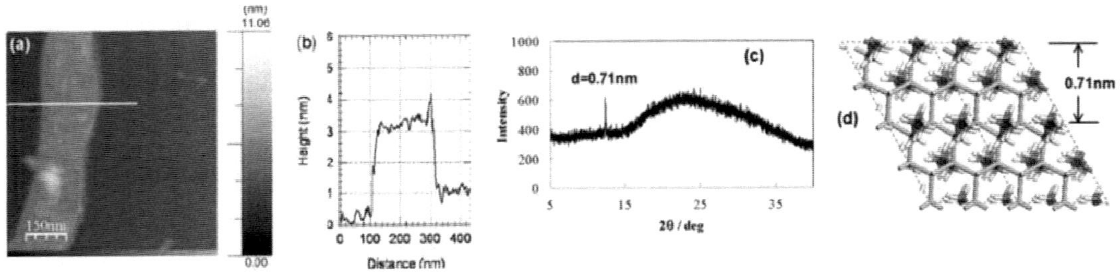

Figure 6. (a) Noncontact mode AFM image of $Si_6H_3(C_6H_{13})_3$, and (b) its line profile taken along the white line in panel a. XRD patterns of (b) $Si_6H_3(C_6H_{13})_3$. (c) Top view of the atomic arrangement of $Si_6H_3(C_6H_{13})_3$ capped with hexane. The dashed line indicates the unit cell. Reproduced from reference [65]

Further results obtained by the authors [65], showed that the two-dimensional silicon sublattice is maintained in the prepared silicon sheet. From the XRD data, the authors suggested that the peak observed at 12.4° indicates molecular-scale periodicity of 0.71 nm, which consists of silicon-hexane composites.[65] The same value (0.71 nm) was calculated from the proposed structural model (Figure 6c) for the periodicity of hexyl groups on the sheet surface.[17,65] However this procedure leads to SiNSs with capping oxygen atoms. The sheet was stabilized though termination of hexane groups pointing out of the layer plane with (111) plane structure of bulk silicon since it was prepared from layered polysilane Si_6H_6.

3.2. Grignard reagent

Another method for the formation of Si-C bond in layered polysilane Si_6H_6 was recently reported by Sugiyama et al.[66] The authors reported the synthesis of oxygen free, phenyl-modified organosilicon nanosheets with atomic thickness by the reaction of layered polysilane [Si_6H_6] with phenyl magnesium bromide [PhMgBr] (Scheme 2).

(h)

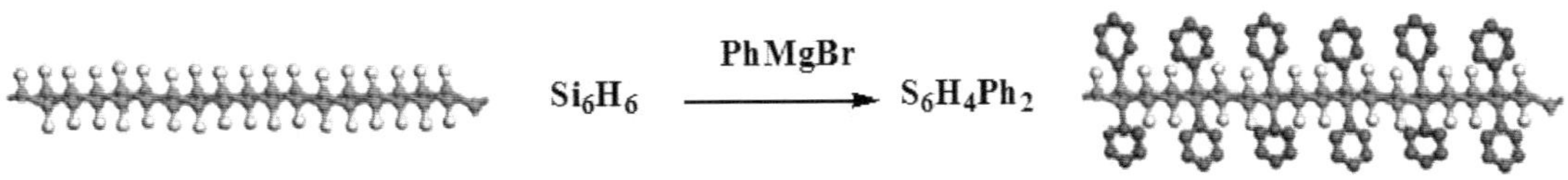

Scheme 2: Reaction of Si_6H_6 with phenyl magnesium bromide [PhMgBr]

The authors suggested that the phenyl group was successfully bonded to the silicon sublattice based on the fact that the material obtained as a colorless paste was soluble in typical organic solvents such as hexane, chloroform, acetone, or ether.

The authors used FTIR and 1H NMR spectroscopy to assign the formula $Si_6H_4Ph_2$ to the obtained SiNSs. The IR spectrum shows vibration bands at 1150 and 1410 cm^{-1} corresponding to the Si-Ph bond and at 1700-2000 cm^{-1} corresponding to aromatic $C = C$ bond indicating that the organic molecules are covalently attached to the silicon surface.1H NMR spectroscopy of $Si_6H_4Ph_2$ also supported results of IR spectroscopy, showing two resonances at 2.5-3.0 and 7.1-7.7 ppm corresponding to those of Si-H and Si-Ph, respectively.

XANES revealed two signals at 1841 and 1844 eV derived from Si-Si and Si-Ph bonds respectively. No signals of Si-O bonds were observed, even in measurements performed after storing the samples in air for one day [66]. The sheet was found to be stable toward oxidation and hydrolysis even with terminating hydrogen atoms on the surface, thus the material could be handled easily in air. Such increased stability of the material by the introduction of Si-C bonds is supported by previous reports on porous silicon[64a] and silicon nanowires (SiNWs).[67]

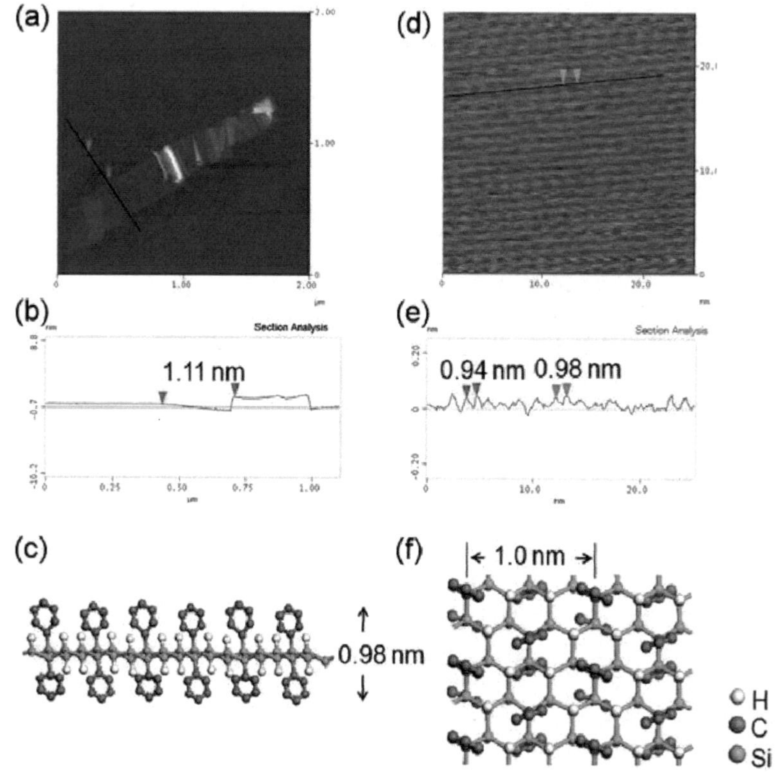

Figure 7. (a), AFM image of $[Si_6H_4Ph_2]$ in contact mode. (b) Line profile along the black line in (a). (c) Side view of the model structure for $[Si_6H_4Ph_2]$. (d) Atomically resolved AFM image of the surface of $[Si_6H_4Ph_2]$.(e) Line profile along the black line in (d). (f) Top view of the model structure for $[Si_6H_4Ph_2]$. Reproduced from reference [66]

The photoluminescence spectrum of $[Si_6H_4Ph_2]$ in 1,4-dioxane using excitation at 350 nm, shows a peak at 415 nm (3.0 eV) and in solid film shows four peaks at 415, 440, 465 (sh), and 505 nm (sh). The photoluminescence indicates the nature of the monolayer sheet, reflecting exfoliation into individual organosilicon nanosheets in an organic solvent and shows similar properties as one-dimensional polysilane.[68] Atomic force microscopy (AFM) and atomically resolved AFM revealed a completely flat plane surface and a periodic structure of the phenyl groups on the silicon surface (Figure 7). The thickness of the sheet was 1.11nm which is close to the thickness (0.98 nm) calculated by using the force field in the program Discover on the basis of its atomic architecture (Figure 7c). Atomically resolved AFM images of an individual sheet show an arrangement of atom-like dots on the top face with closest distance between the dots of $0.96 \pm 0.02nm$. The periodicity of phenyl groups on the model surface (Figure 7f), was estimated to be 1.0 nm, which is in good agreement with the closest distance between the dots in the AFM image. The approximate 1.0 nm periodicity was also confirmed by XRD.

3.3. Reaction with amine groups

Si atoms can easily accept nucleophilic attack of a lone pair from the amine N atoms because the Si-H bonds in Si_6H_6 are similar to those in silyl hydride, that is, the bonds are polarized as $Si^{\delta+} - H^{\delta-}$. It is known that ammonia can be adsorbed on a Si surface and can then react with dangling bond sites without major disruption to the Si surface structure.[69] Formation of $Si - N$ bonds upon reaction of n-alkylamines with chlorinated or hydrogenated Si surface have been reported.[70-72] Based on these studies, an approach to the synthesis of oxygen-free silicon nanosheets covered with amines using Si_6H_6 as a starting material have been recently reported by Okamoto et al.[73] Reaction of Si_6H_6 with n-Decylamine was monitored in chloroform at 60°C under nitrogen atmosphere for approximately 1 h (Scheme 3). Reaction progress is evidenced by a change in color and flotation of Si_6H_6. A similar change was observed when n-butylamine was used instead of n-decylamine, but not for di- or tri-n-butylamine.[73] Bubble formation observed just after pouring the amine onto Si_6H_6 implies that hydrogen is generated as a result of the reaction.

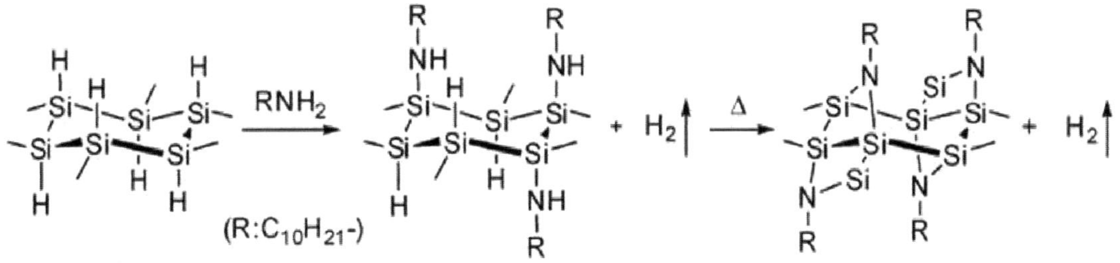

Scheme 3: Proposed Reaction Scheme for the production of $C_{10} - Si_n$ from Si_6H_6 and n-Decylamine[73].

The authors suggested two possible outcomes concerning the binding mode in this surface modification; the amine either reacts with Si_6H_6 or attaches to the surface of the silicon layers. [73,74] In the case of reaction (Scheme 3), some Si-H bonds react with amine molecules to form $Si - NH - R$ linkages (R: decyl group) which subsequently react through its $N - H$ bond with another $Si - H$ bond on $Si_6H_6$to form a Si-NR-Si linkage. The final product contains both $Si - NH - R$ and $Si - NR - Si$ linkages in variable proportions, depending on the reaction conditions.

FTIR of the product (Figure 8) shows bands from 2800 to 3000 cm^{-1} and from 1350 to 1450 cm^{-1} attributed to the alkyl chain of n-decylamine. There is no band around 2100 cm-1, indicating $Si - H$ stretching vibrations and no $N - H$ stretching around 3300 cm^{-1}. Asymmetric $Si - N - Si$ stretching was also observed as weak peak at 930 cm^{-1}. [74] These initial analyzes support the reaction procedure of the amine with Si_6H_6 according to the proposed mechanism (Scheme 3).

XANES spectra (Figure 8) also revealed Si-N linkage in the layered polysilane and did not contain $Si - O$ linkages. These results indicate that n-decylamine is not merely attached to

Si_6H_6 but has actually reacted with Si_6H_6 . Thermogravimetric analysis (TGA) of the product revealed that alkyl-chain moieties remained connected to the Si layers up to 400°C and suggested that the silicon layers in $(C_{10} - Si_n)$ were densely covered with n-decylamine residues with an estimated ca. 0.7 mol of residue per mole of Si atoms in the reaction product.[73]

A small amount of $C_{10} - Si_n$ is dispersible in chloroform and exfoliates to form SiNSs with lateral dimensions in the range of 12 mm, as shown by AFM (Figure 8d). SiNSs have flat and smooth surfaces, due to dense coverage of n-decylamine.[73] The thickness of the sheets was found to be 7.5 nm (Figure 8e) 10-20 times larger than that of a siloxene nanosheet [57] or than that of a Mg-doped silicon nanosheet [58] due to the long alkyl chains connected to the surface of the $C_{10} - Si_n$ nanosheet on both sides.

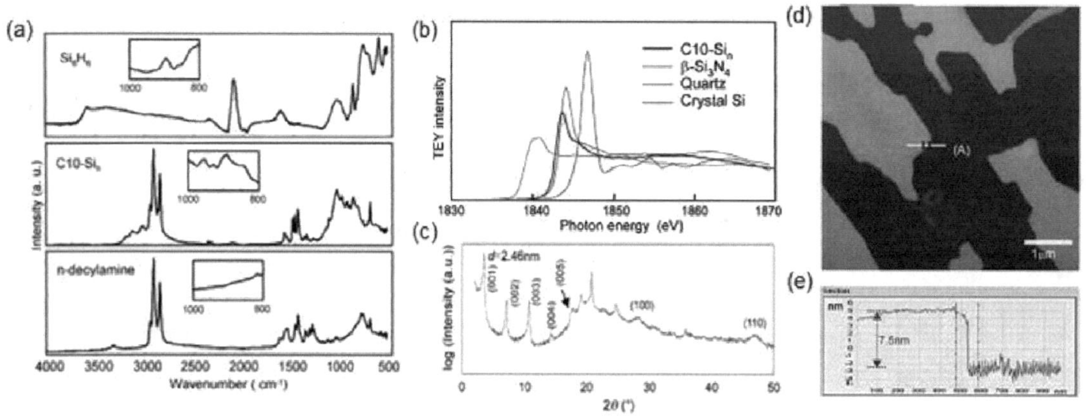

Figure 8. (a) FTIR spectra of Si_6H_6, n-decylamine, and $C_{10} - Si_n$. The insets show enlarged spectra in the range from 800 to 1000 cm^{-1}. (b) Comparison of XANES spectra of C10-Sin with those of quartz, crystal silicon, and -Si_3N_4 obtained using the TEY mode in a vacuum. (c) XRD pattern of $C_{10} - Si_n$ in the agglomerated state. (d) AFM phase image of a $C_{10} - Si_n$ nanosheet on a HOPG plate obtained in tapping mode. (e) Line profile measured along the white line denoted as (A) in panel a. Reproduced from reference [73]

XRD show peaks of puckered silicon sheets ((100) and (110) planes) with no peak corresponding to the (001) planes of Si_6H_6. This result indicates that Si_6H_6 was transformed into $C_{10} - Si_n$ through reaction with n-decylamine, and the crystalline sheet structure was preserved after the reaction. In the low-angle region a peak derived from (001) planes appear with interlayer distance estimated to be 2.46 nm close to 2.54 nm, which is the expanded interlayer distance of siloxene due to intercalation of n-decylamine.[45]

The silicon sheets prepared by hydrosilylation, Grignard reaction or amination of Si_6H_6 were stabilized though termination of organic groups pointing out of the layer plane with (111) plane structure of bulk silicon since they were prepared from layered polysilane (Si_6H_6).

4. Conclusions and Future Perspectives

Silicon-based nanotechnology is highly promising since it is compatible with conventional silicon microtechnology. The structural feature of silicon nanomaterials is their large specific surface areas, which are due to their size and shape on the nanometer scale. Here, we reviewed chemical synthesis methods of two dimensional silicon nanomaterials (SiNSs). We have described the developed soft synthetic method for the preparation of SiNSs by topochemical reaction of $CaSi_2$, modified $CaSi_2$ and by the chemical surface modification of Si_6H_6. Stacked silicon layer scan be exfoliated from $CaSi_2$ to provide SiNSs by using surfactants, compositional modification by Mg doping, and surface modification of Si_6H_6 with organic compounds using hydrosilylation, Grignard reaction and amination reaction. SiNSs maintain their characteristic 2D crystalline structure and stabilized through termination with organic groups pointing out of the layer plane with (111) plane structure of bulk silicon since they were prepared from layered polysilane (Si_6H_6).

Recently, we synthesized extended silicon sheets by simple room temperature procedure called redox assisted chemical exfoliation of calcium disilicide ($CaSi_2$). The data clearly show that the synthesized Si sheets do not correspond to the (111) oriented layers of Si diamond lattice; instead they have hexagonal graphitic-like structure resembling that of silicene. Silicon in the form of silicene may form a new avenue for an old player in the electronic industry, which may extend the lifetime of Moores law.

The application of this new synthetic approach is expected to be extended to the use of other systems and theoretical calculations should be performed to predict the product properties. SiNSs are expected to be developed for applications in various fields that can take advantage of their specific characteristics.

References

[1] Teo B K and Sun X H 2007 *Chem. Rev.* **107** 1454
[2] Canham L T 1990 *Appl. Phys. Lett.* **57** 1046
[3] Lehmann V and Gçssele U 1991 *Appl. Phys.* Lett. **58** 856
[4] Bley R A and Kauzlarich S M 1996 *J. Am. Chem. Soc.* **118** 12461
[5] Tilley R D, Warner J H, Yamamoto K, Matsui I and Fujimori H 2005 *Chem. Commun.* **1833**
[6] Morales A M and Lieber C M 1998 *Science* **279**, 208
[7] Molnar W, Lugstein A, Pongratz P, Auner N, Bauch C and Bertagnolli E 2010 *Nano Lett.* **10** 3957
[8] Sha J, Niu J, Ma X, Xu J, Zhang X, Yang Q and Yang D 2002 *Adv. Mater.* **14** 1219
[9] De Crescenzi M, Castrucci P, Scarselli M, Diociaiuti M, Chaudhari P S, Balasubramanian C, Bhave T M and Bhoraskar S V 2005 *Appl. Phys. Lett.* **86** 231901
[10] Kara A, Enriquez H, Seitsonen A P, Lew Yan Voon L C, Vizzini S, Aufray B and Oughaddou H 2012 *Surf. Sci. Rep.* **67** 1
[11] Schaak R E and Mallouk T E 2002 *Chem. Mater.* **14** 1455
[12] Sasaki T 2007 *J. Ceram. Soc. Jpn.,* **115** 9
[13] Miyamoto N, Yamamoto H, Kazuyuki Kuroda R. K. 2002 *Chem. Commun.* **2378**
[14] Kai K, Yoshida Y, Kageyama H, Saito G, Ishigaki T, Furukawa Y and Kawamata J 2008 *J. Am. Chem. Soc.* **130** 15938
[15] Tae E L, Lee K E, Jeong J S and Yoon K B 2008 *J. Am. Chem. Soc.* **130** 6534
[16] Teo B K and Sun X H 2007 *Chem. Rev.* **107** 1454.
[17] Okamoto H, Sugiyama Y and Nakano H 2011 *Chem. Eur. J.* **17** 9864
[18] Jutzi P and Schubert U Silicon Chemistry: From the Atom to Extended Systems. Copyright © 2003 *Wiley-VCH Verlag GmbH and Co. KGaA, Weinheim,* pp 506
[19] Takeda K and Shiraishi K 1994 *Phys. Rev. B* **50** 14916
[20] Léandri C, Oughaddou H, Aufray B, Gay J. M, Le Lay G, Ranguis A and Garreau Y 2007 *Surf. Sci.* **601** 262
[21] Guzman-Verri G G and Lew Yan Voon L C 2007 *Phys. Rev. B* **76** 075131. Léandri C, Le Lay G, Aufray B, Girardeaux C, Avila J, Davila M E, Asensio M C, Ottaviani C and Cricenti A 2005 *Surf. Sci.* **574** L9
[22] Le Lay G, Aufray B, Léandri C, Oughaddou H, Bibérian J-P, De Padova P, Dávila M E, Ealet B and Kara A 2009 *Appl. Surf. Sci.* **256** 524
[23] Kara A, Léandri C, Dávila M E, De Padova P, Ealet B, Oughaddou H, Aufray B and Le Lay G 2009 *J. Supercond. Novel Magn.* **22** 259
[24] Aufray B, Kara A, Vizzini S, Oughaddou H, Léandri C, Ealet B and Le Lay G 2010 *Appl. Phys. Lett.* **96** 183102
[25] Lalmi B, Oughaddou H, Enriquez H, Kara A, Vizzini S, Ealet B and Aufray B 2010 *Appl. Phys. Lett.* **97** 223
[26] Majzik Z, Tchalala M R, Švec M, Hapala P, Enriquez H, Kara A, Mayne A J, Dujardin G, Jelínek P and Oughaddou H 2013 *J. Phys.: Condens. Matter.* **25** 225301
[27] Tchalala M R, Enriquez H, Mayne A J, Kara A, Roth S, Silly M G, Bendounan A, Sirotti F, Greber T, Aufray B, Dujardin G, Ait Ali M and Oughaddou H 2013 *Appl. Phys. Lett.* **102** 083107

[28] Meng L, Wang Y, Zhang L, Du S, Wu R, Li L, Zhang Y, Li G, Zhou H, Hofer W A and Gao H-J 2013 *Nano Lett.* **13** 685
[29] Fleurence A, Friedlein R, Ozaki T, Kawai H, Wang Y and Yamada-Takamura Y 2012 *Phys. Rev. Lett.* **108** 245501
[30] Houssa M, Pourtois G, Afanassev V V and Stesmans A 2010 *Appl. Phys. Lett.* **97** 112106
[31] Dzade N Y, Obodo K O, Adjokatse S K, Ashu A C, Amankwah E, Atiso C D, Bello A A, Igumbor E, Nzabarinda S B, Obodo J T, Ogbuu A O, Femi O E, Udeigwe J O and Waghmare U V 2010 *J. Phys. Condens. Matter* **22** 375502
[32] Yamanaka S 2010 *Dalton Trans.* **39** 1901
[33] Mayer I P, Banks E and Post B 1962 *J. Phys. Chem.* **66** 693
[34] Tchalala M R, Ait Ali M, Enriquez H, Kara A, Lachgar A, Yagoubi S, Foy E, Vega E, Bendounan A, Silly M G, Sirotti F, Nitshe S, Chaudanson D, Jamgotchian H, Aufray B, Mayne A J, Dujardin G and Oughaddou H 2013 *J. Phys.: Condens. Matter* **25** 44200
[35] Gärtner S and Korber N 2011 Struct Bond **140** 25
[36] Kauzlarich S M Chemistry, Structure and Bonding of Zintl Phases and Ions, *Wiley-VCH Publishers, New York* 1996
[37] Schäfer R 1985 *Annu. Rev. Mater. Sci.* **15** 1
[38] Schäfer H, Eisenmann B and Müller W 1973 *Angew. Chem. Int. Ed. Engl.* **12** 694
[39] McMillan P F, Gryko J, Bull C, Arledge R, Kenyon A J and Cressey B A 2005 *J. Solid State Chem.* **178** 937
[40] Kasper J S, Hagenmuller P, Pouchard M and Cros C 1965 *Science* **50** 1713
[41] San-Miguel A and Toulemonde P 2005 *High Pressure Res.* **25** 159
[42] Ramachandran G K, Dong J, Diefenbacher J, Gryko J, Marzke R F, Sankey O F and McMillan P F 1999 *J. Solid State Chem.* **145** 716
[43] Böhm J and Hassel O 1927 *anorg. Z. allg. Chem.* **160** 152
[44] Wallbaum H J 1944 *Naturwissenschaften* **32** 76
[45] Weiss A, Beil G and Meyer H Z 1979 *Naturforsch.* **35b** 25
[46] Bonitz E 1961 *Chem. Ber.* **94** 220
[47] Kautsky H 1952 Z. *Naturforsch.* **7b** 174
[48] Bonitz E 1966 *Angew. Chem. Int. Ed. Engl.* **5** 462
[49] Brandt M S, Vogg G and Stutzmann M in Silicon Chemistry: From the Atom to Extended Systems (Eds.: Jutzi P, Schubert U), *Wiley-VCH, Weinheim*, 2003, pp. 194–213
[50] Kautsky H 1921 *Z. anorg. Chem.* **117** 209
[51] Hengge E 1967 *Fortschr. Chem. Forsch.* **9** 145
[52] Dettlaff-Weglikowska U, Hönle W, Molassioti-Dohms A, Finkbeiner S and Weber J 1997 *Phys. Rev. B* **56** 13132
[53] Dahn J R, Way B M, Fuller E and Tse J S 1993 *Phys. Rev. B* **48** 17872
[54] Sugiyama Y, Okamoto H and Nakano H 2010 *Chem. Lett.* **39** 938
[55] Yamanaka S, Matsu-ura H and Ishikawa M 1996 *Mater. Res. Bull.* **31** 307
[56] Nishimura K, Nagao Y, Yamanaka S and Matsu-ura H 1996 Jpn. *J. Appl. Phys.* **35** L293
[57] Nakano H, Ishii M and H. Nakamura 2005 *Chem. Commun.* 2945
[58] Nakano H, Mitsuoka T, Harada M, Horibuchi K, Nozaki H, Takahashi N, Nonaka T, Seno Y and Nakamura H 2006 Angew. Chem. 118 6451-6454, 2006 *Angew. Chem. Int. Ed.* **45** 6303
[59] Enriquez H, Vizzini S, Kara A, Lalmi B and Oughaddou H 2012 *J. Phys.: Condens. Matter.* **24** 314211
[60] Arafune R, Lin C-L, Kawahara K, Tsukahara N, Minamitani E, Kim Y, Takagi N and Kawai M 2013 *Surf. Sci.* **608** 297
[61] Feng B, Ding Z, Meng S, Yao Y, He X, Cheng P, Chen L and Wu K 2012 *Nano Lett.* **12** 3507
[62] Cahangirov S, Topsakal M, Aktürk E, ?ahin H and Ciraci S 2009 *Phys. Rev. Lett.* **102** 236804
[63] Himpsel F J, Hollinger G and Pollak R A 1983 *Phys. Rev. B* **28**, 7014
[64] (a) Buriak, J M 1999 *Chem. Commun.* 1051. (b) Buriak, J M 2002 *Chem. Rev.* **102** 1271
[65] Nakano H, Nakano M, Nakanishi K, Tanaka D, Sugiyama Y, Ikuno T, Okamoto H and Ohta T. 2012 *J. Am. Chem. Soc.* **134** 5452
[66] Sugiyama Y, Okamoto H, Mitsuoka T, Morikawa T, Nakanishi K, Ohta T and Nakano H 2010 *J. Am. Chem. Soc.* **132** 5946
[67] Haick H, Hurley P T, Hochbaum A I, Yang P and Lewis N S 2006 *J. Am. Chem. Soc.* **128** 8990
[68] Miller R D and Michl J 1989 *Chem. ReV.* **89** 1359
[69] Waltenburg H N and Yates J T 1995 *J. Chem. ReV.* **95** 1589
[70] Bergerson W F, Mulder J A, Hsung R P and Zhu X-Y 1999 *J. Am.Chem. Soc.* **121** 454–455
[71] Rogozhina E, Belomoin G, Smith A, Abuhassan L, Barry N, Akcakir O, Braun P V and Nayfeh M H 2001 *Appl. Phys. Lett.* **78** 3711

[72] Liao Y-C and Roberts J T 2006 *J. Am. Chem. Soc.* **128** 9061
[73] Okamoto H, Kumai Y, Sugiyama Y, Mitsuoka T, Nakanishi K, Ohta T, Nozaki H, Yamaguchi S, Shirai S and Nakano H 2010 *J. Am. Chem. Soc.* **132** 2710.
[74] Socrates G Infrared and Raman Characteristic Group Frequencies; Table and Charts, 3rd ed.; *Wiley: New York*, 2001

AUTHOR INDEX

Institute of Physics
Dirac House, Temple Back
Bristol BS1 6BE UK

CURRAN ASSOCIATES INC
proceedings.com